I0899345

Pharmacology and
Biochemistry of
Psychiatric Disorders

Pharmacology and Biochemistry of Psychiatric Disorders

A. Richard Green
MRC Unit & University Department of Clinical Pharmacology
Radcliffe Infirmary, Oxford

and

David W. Costain
University Department of Psychiatry
Littlemore Hospital, Oxford

A Wiley–Interscience Publication

RELEASED

JOHN WILEY & SONS
Chichester · New York · Brisbane · Toronto

Copyright © 1981 by John Wiley & Sons Ltd.

All rights reserved.

No part of this book may be reproduced by any means, nor
transmitted, nor translated into a machine language
without the written permission of the publisher.

British Library Cataloguing in Publication Data:

Green, A. Richard
   Pharmacology and biochemistry of psychiatric
   disorders.
   1. Neurochemistry
   2. Neuropharmacology
   I. Title   II. Costain, David W.

   ISBN 0 471 09998 8 (cloth)
   ISBN 0 471 10000 5 (paper)

Typeset by Pintail Studios Ltd., Ringwood, Hampshire.
Printed in Great Britain by The Pitman Press Ltd., Bath, Avon.

Contents

Preface xiii

Chapter 1   INTRODUCTION – STRATEGIES IN PSYCHO-
            PHARMACOLOGY RESEARCH   1
1.1   The problems of psychiatric assessment   1
1.2   The problems of investigating clinical biochemical changes . .   3
1.3   The problems of investigating the pharmacology of psychoactive
      drugs   5
1.4   The value of psychopharmacological research   7

Chapter 2   BIOCHEMISTRY OF THE NEUROTRANSMITTERS .   8

      PART 1   5-HYDROXYTRYPTAMINE
2.1   Isolation and identification   8
2.2   Distribution   8
2.3   Synthesis   9
2.4   Metabolism   14
2.5   Control of synthesis, metabolism, and function   16

      PART 2   THE CATECHOLAMINES
2.6   Isolation and identification   18
2.7   Distribution   18
2.8   Synthesis   19
2.9   Metabolism   22
2.10 Control of synthesis and release   23
2.11 Presynaptic receptors   23
2.12 Turnover of neurotransmitters   25
2.13 Dopamine- and noradrenaline-sensitive adenylate cyclase . .   25
2.14 Ligand–receptor binding techniques   27

      PART 3   ACETYLCHOLINE
2.15 Introduction   30
2.16 Synthesis   31
2.17 Metabolism   32
2.18 The acetylcholine receptor   33
2.19 Cholinergic agonists and antagonists   34
2.20 Quantal release   34
2.21 Criteria for a neurotransmitter   35

PART 4   OTHER SMALL MOLECULES

2.22   γ-Aminobutyric acid (GABA)   35
2.23   L-Glutamic acid   38
2.24   Glycine   38
2.25   Histamine   38

PART 5   PEPTIDE TRANSMITTERS

2.26   General introduction   39
2.27   Opiate peptides – introduction   41
2.28   Historical aspects of opiate peptides   41
2.29   The enkephalins   42
2.30   β-Endorphin   42
2.31   Multiple enkephalin receptors   43
2.32   Distribution of enkephalins   43
2.33   Synthesis, degradation, and release of enkephalins . . .   44
2.34   Thyrotropin releasing hormone (TRH)   44
2.35   Substance P   46
2.36   MIF or PLG   47
2.37   The corticotrophin-related hormones   48
2.38   Vasopressin   48

Chapter 3   DEPRESSION AND MANIA   49

PART 1   PSYCHIATRIC ASPECTS OF DEPRESSION

3.1   General introduction   49
3.2   Epidemiology   49
3.3   Clinical features   50
3.4   Classification   51
3.5   Unipolar versus bipolar   53
3.6   Primary versus secondary   54
3.7   Type A versus type B   54
3.8   Classification according to family history   55
3.9   Quantification of ratings of depression   56

PART 2   BIOCHEMICAL ASPECTS OF DEPRESSION

3.10   General introduction   57
3.11   Historical aspects   58
3.12   Cerebrospinal fluid indoleamines   60
3.13   Post-mortem brain studies   63
3.14   Plasma tryptophan   64
3.15   Oral contraceptives, tryptophan metabolism, and depression .   65
3.16   Urinary 5-HIAA   67
3.17   The blood platelet   68
3.18   Catecholamine metabolites and cerebrospinal fluid   68
3.19   Plasma, platelet, and urinary catecholamine studies   69
3.20   In vivo studies on neurotransmitter function   70

PART 3   THE PHARMACOLOGY OF DEPRESSION

3.21	Historical aspects	71
3.22	Monoamine oxidase inhibitors	71
3.23	L-Tryptophan plus a MAOI	74
3.24	L-Tryptophan	74
3.25	Tricyclic antidepressants	75
3.26	Tricyclic binding to brain and platelet	82
3.27	Dosage of tricyclics	82
3.28	Lithium	83
3.29	Electroconvulsive therapy	83
3.30	General conclusions	87

PART 4   MANIA

3.31	Psychiatric aspects	88
3.32	Biochemistry	89
3.33	Pharmacology – introduction	89
3.34	Lithium – historical aspects	90
3.35	Measurement and therapeutic plasma levels of lithium	90
3.36	Pharmacology of lithium	91
3.37	Other pharmacological approaches	92

Chapter 4   ANXIETY   93

PART 1   PSYCHIATRIC ASPECTS

4.1	Symptomatology	93
4.2	Classification	95
4.3	Obsessional disorders	95

PART 2   PHARMACOLOGY

4.4	General introduction	96
4.5	The benzodiazepines	96
4.6	Brain benzodiazepine receptors and GABA modulin	98
4.7	Is there an endogenous benzodiazepine?	99
4.8	Is the anxiolytic action associated with GABA?	100
4.9	Benzodiazepine metabolism and kinetics	101
4.10	Dependence on benzodiazepines	102
4.11	Benzodiazepines and seizure disorders	102
4.12	Barbiturates	102
4.13	β-Adrenoceptor antagonists	102
4.14	Meprobamate	102
4.15	Tricyclic antidepressants	102
4.16	Monoamine oxidase inhibitors	103

Chapter 5   SCHIZOPHRENIA   104

PART 1   PSYCHIATRIC ASPECTS OF SCHIZOPHRENIA

5.1	Introduction	104

5.2   Clinical features 104
5.3   Schizo-affective disorders 106
5.4   Diagnosis 106
5.5   Standardized diagnostic schedules 106
5.6   Outcome 108
5.7   Assessment of outcome 109
5.8   Aetiological theories 110

PART 2   THE BIOCHEMISTRY OF SCHIZOPHRENIA

5.9   Endogenous psychotogens 110
5.10 Monoamine neurotransmitter systems 111
5.11 The possible involvement of a virus 114

PART 3   THE PHARMACOLOGY OF SCHIZOPHRENIA

5.12 Introduction 114
5.13 The neuroleptic drugs 115
5.14 Depot neuroleptics 116
5.15 Action of neuroleptics on dopamine metabolism and behaviour   118
5.16 Dopamine-sensitive adenylate cyclase 119
5.17 Ligand binding studies 121
5.18 The mesolimbic forebrain 124
5.19 Tardive dyskinesia 125
5.20 The effect of long term neuroleptic administration to rats . . . 126
5.21 Dopamine antagonism as an explanation for the antipsychotic
      action of neuroleptics 127

Chapter 6   ALZHEIMER'S DISEASE AND SENILE DEMENTIA   . 129

PART 1   CLINICAL ASPECTS

6.1   Introduction 129
6.2   Prevalence 129
6.3   Clinical features 129

PART 2   BIOCHEMISTRY

6.4   Pathological changes 130
6.5   Brain catecholamines 131
6.6   Brain 5-hydroxytryptamine 133
6.7   Brain monoamine enzymes 133
6.8   Brain GABA 133
6.9   Brain cholinergic systems 134

PART 3   PHARMACOLOGY

6.10 Introduction 136
6.11 Improvement of cerebral blood flow and oxygenation 137
6.12 Cholinomimetic drugs 137
6.13 Future approaches 138

Chapter 7   HUNTINGTON'S CHOREA **139**

PART 1   CLINICAL ASPECTS
7.1   Clinical features   139

PART 2   BIOCHEMISTRY
7.2   Pathological changes   139
7.3   Brain GABA   140
7.4   Brain catecholamines   141
7.5   Brain acetylcholine   142
7.6   Brain 5-HT   142
7.7   General conclusions   143

PART 3   PHARMACOLOGY
7.8   Possible therapeutic approaches   143

Chapter 8   PARKINSON'S DISEASE **144**

PART 1   CLINICAL ASPECTS
8.1   Historical aspects   144
8.2   Clinical features   144
8.3   Prevalence   145
8.4   Natural history   145
8.5   The 'on–off' phenomenon   145

PART 2   BIOCHEMISTRY
8.6   Introduction   146
8.7   Dopamine in the substantia nigra   146
8.8   Cerebrospinal fluid dopamine metabolite concentrations   . . .   147
8.9   Brain 5-HT metabolism   147
8.10 Brain GABA   148

PART 3   PHARMACOLOGY
8.11 Introduction   149
8.12 L-Dopa therapy   149
8.13 Use of peripheral decarboxylase inhibitors with L-dopa   . . .   150
8.14 L-Dopa plus pyridoxine administration   151
8.15 L-Dopa plus a MAO inhibitor   151
8.16 The value of L-dopa   152
8.17 Bromocriptine   152
8.18 Future approaches to increase dopamine function   153
8.19 Amantadine   153
8.20 Anticholinergics   153
8.21 Electroconvulsive therapy   153
8.22 PLG   154
8.23 GABA mimetics   154
8.24 Conclusion   154

Chapter 9   DRUG DEPENDENCY **155**
9.1   General introduction: social aspects, tolerance, and dependence   155

PART 1   ALCOHOL

9.2   Social aspects 156
9.3   Physical effects 157
9.4   Clinical aspects 157
9.5   Metabolism 158
9.6   Mechanism of addiction 158

PART 2   CANNABIS OR MARIJUANA

9.7   Physical and psychological effects 160
9.8   Pharmacological effects 160

PART 3   OPIATES

9.9   Historical aspects 161
9.10   Physical and psychological effects 162
9.11   Opiate dependence and withdrawal 163

PART 4   HALLUCINOGENS

9.12   Historical and social aspects 164
9.13   Physical and psychological effects of LSD 164
9.14   Pharmacological effects of LSD 165
9.15   Other hallucinogenic compounds 166

PART 5   PHENCYCLIDINE

9.16   Physical and psychological effects 166
9.17   Pharmacological aspects 167

PART 6   AMPHETAMINES

9.18   Physical and psychological effects 167
9.19   Pharmacological aspects 168

PART 7   COCAINE

9.20   Physical, psychological, and pharmacological effects 169

PART 8   BENZODIAZEPINES

9.21   Problems of dependence 170

PART 9   BARBITURATES

9.22   Physical and psychological effects 170
9.23   Pharmacological aspects 171
9.24   Metabolism 172

PART 10   NICOTINE (TOBACCO)

9.25   Physical and psychological effects 172
9.26   Mechanism of dependence 172

**Appendix 1   COMMON ABBREVIATIONS USED IN
NEUROPHARMACOLOGY** **174**

Appendix 2   DRUGS CITED IN TEXT WITH SOME OF THEIR
              COMMON TRADE NAMES 176

Appendix 3   DRUG TRADE NAMES WITH THEIR NON-
              PROPRIETARY NAMES 179

Appendix 4   NEUROTRANSMITTERS; METABOLIC INHIBITORS,
              AGONISTS AND ANTAGONISTS 182

Glossary   184

References   191

Index   205

Preface

It is apparent that there is no book which is suitable for recommending to undergraduate and graduate students studying central nervous system pharmacology or to psychiatrists and others studying biochemical and pharmacological aspects of psychiatric disorders. Instead a long reading list of publications is required which fails to provide an overall perspective.

There are several excellent books dealing with either more basic aspects of neuropharmacology (particularly Cooper, Bloom and Roth, 1978) or therapeutic aspects of neuropharmacology (e.g. Barchas *et al.*, 1977). The aim of this book is to provide an overview of the current state of knowledge of the pharmacology and biochemistry of psychiatric disorders in a form accessible to readers approaching the subject from different disciplines. We have therefore examined the experimental pharmacological data on drugs used in psychiatry and discussed whether they can be linked with clinical and biochemical findings. We have avoided trying to 'make a story' in order to reach a satisfying conclusion but rather reviewed the major information: it is still possible to present information to support the notion that there is decreased monoamine function in depression; however, current experimental data do not support the idea very strongly. We have therefore presented the results that seem to be important. Readers having been given some of the historical data and more recent research will then, we hope, be in a better position to evaluate the current research.

However, we have included two chapters that do not directly reflect these aims. The first is a chapter on the basic biochemistry of neurotransmitters as some knowledge is necessary for understanding both the pharmacology and the clinical biochemistry. The second is a chapter on Parkinson's disease, which is included because it serves as a paradigm for a condition caused by abnormalities in a brain neurotransmitter system, and where the pharmacological treatment has developed from biochemical studies.

The sections on clinical aspects have been included for two reasons. They allow workers with little clinical experience (pharmacologists, experimental psychologists, or preclinical students) to get some 'feel' for the disorder for which the drugs are given. They also allow us to point to problems associated with clinical neuropharmacological research. These sections are not intended to be comprehensive and in some cases are oversimplified for the sake of clarity, although not to the point of being misleading. For those wishing to pursue clinical aspects further, modern texts have been published in the United States (Freedman, Kaplan and Sadock, 1980), and by workers at the Institute of Psychiatry (Hill, Murray and Thorley, 1979). There will shortly also be one from the University of Oxford (Gelder, Gath and Mayou, 1982).

Not all the clinical sections follow the same format, as the aetiology, pathology, and treatment of different disorders may be concerned to different extents with the pharmacology. We have nevertheless tried to maintain as much uniformity as possible.

Papers have been cited for three reasons: that they are major papers (perhaps for historical reasons), that they are good reviews, or that they report work too recent to have been fully reviewed elsewhere. This system may be unfair to some workers in that it fails to give them credit for important work, but it prevents the reference list becoming unwieldy.

We have included several appendices, in an attempt to provide easy access to some basic facts. The glossary provides a dictionary of specialist terminology which may be encountered in this and other publications.

Not surprisingly, the book reflects our own views, and indeed prejudices, on current important areas of research. In a few years some sections will be out of date. We hope that this will reflect advances in knowledge rather than a failure on our part to identify important trends.

We thank the various publishers and authors who have generously allowed us to reproduce data from their publications. It is a pleasure also to thank Fiona Teddy for her typing of the manuscript. Finally we should like to acknowledge the invaluable comments of many of our colleagues, particularly Dr Philip J. Cowen and Dr David J. Nutt. We have not always followed their advice and remaining errors are the responsibility of the authors.

A. RICHARD GREEN
DAVID W. COSTAIN

Chapter 1

Introduction – Strategies in Psychopharmacology Research

1.1 The problems of psychiatric assessment

A major problem for research in psychiatry, and particularly for psychopharmacological research, is that of accurate diagnosis; that is the delineation of specific conditions or illnesses. The problem of classification is a continuing one for psychiatry, and new systems are regularly introduced as offering advances over existing ones. Although most psychiatrists are agreed on the typical characteristics of schizophrenia, depression, and other disorders, devising a classification which meets even the very basic requirements of having categories which are mutually exclusive and jointly exhaustive has proved elusive, atypical cases being difficult to categorise. When one also includes the requirement that a classification should provide implications for both treatment and outcome, the position becomes fraught with difficulties.

Much of the research and controversy in psychiatry has been concerned with this exact clinical delineation (see Section 3.4) and whilst there has been some validation of the distinction between the major illnesses by differential responses to treatment, subdivisions of these illnesses and the placing of atypical cases, have been less successful.

Such difficulties have led some people to question the validity of the concept of psychiatric illness. There is insufficient space here to do justice to arguments which could easily fill the whole of this book, but it is not to avoid the issue to suggest that an open mind should be kept on the concept of illness, whilst arguing that a consideration of patterns of distress, and ways of relieving it, is a justifiable goal.

Study of the development of medical classifications of illness shows a change in the basis of the systems. Initially classified on signs and symptoms, as knowledge progresses this system is replaced by a classification based on pathology (usually derived from investigating groups identified on the basis of physical signs) and finally by a system which considers aetiology. This last is the ultimate aim for nosologists, and does not necessarily require a knowledge of specific pathology, as it may be inferred from epidemiological data.

With a few exceptions in the field of mental subnormality, psychiatric illnesses are recognized and classified solely on the basis of symptoms and signs. Unlike many (but not all) physical illnesses there is no objective biochemical or physiological test which can be used to confirm or validate any diagnosis, give any indication of appropriate treatment, or help to predict outcome.

Why is this thought to be a reasonable goal, and why has so much effort been expended on the search for neurochemical changes in psychiatric illness? One answer to the first question is that historically, conditions which were included with psychiatric illness were found to have an organic basis when appropriate technology became available. The prime example of this is General Paralysis of the Insane, caused by the spirochaete of syphilis, but others include some cranial tumours, and metabolic disorders. More rational forms of treatment were then introduced. In answer to the second question it is now known that a variety of hormonal and metabolic disturbances, either 'spontaneous' or drug induced, produce symptoms indistinguishable from psychiatric disorders, and that the drugs effective in treating psychiatric illness have marked actions on central neurotransmitter systems, suggesting that some biological process may be associated with the clinical symptoms. This line of thought, however, leads to an assumption generally made, that more specific delineation of clinical symptomatology improves the likelihood of finding a biochemical marker of pathology.

This is of course justifiable only where there is a one to one relationship between clinical symptoms and aetiology; that is to say that there is a pattern to the symptoms and signs produced by any causal factor, and that these are specific to it. The paradigm of this is infective illness. This is unlikely to be the case in psychiatry since there is evidence that aetiology of the major illnesses is multifactorial, that many symptoms occur in different disorders and single disorders have a wide variety of presentations of symptoms.

That this approach should have produced results in the past is evidence only that *some* conditions have such a one to one relationship, and were therefore selectively identified. It may be that the residual 'pool' of illness has much more complicated aetiological relationships.

Although some of the classificatory systems are described as multifactorial in that they include factors other than clinical symptoms and signs (e.g. duration, physical illness, social functioning, past history, family history), most combine all these factors in a single diagnostic schedule, thereby confusing the contributions of the various factors in the aetiology. An exception is the *Diagnostic and Statistical Manual of Mental Disorders* of the American Psychiatric Association (DSM-III) (Task Force, 1980) which defines five separate axes, but which nevertheless fails to provide separate evaluations of past history, family history, or specific social factors.

The important point for research, however, remains that diagnoses made should be reliable and generally agreed as valid. Both the research diagnostic criteria of Spitzer *et al.* (1975) (which has influenced the DSM-III) and the Present State Examination (PSE) with computerized CATEGO diagnosis of Wing, Cooper and Sartorius (1974; see Section 5.5) meet these criteria and are widely used.

A further problem is that of rating psychiatric illness. This is discussed in the appropriate sections but there are some general points.

The function of these scales is quite distinct from diagnosis. The scales are tools to assess the severity of particular diagnoses and they should measure reliably what they are intended to measure. This is usually assessed when the scales are introduced, by comparison with a general clinical assessment, or with pre-existing scales. It is important to note, however, that they are usually assessed in specific clinical settings, (e.g. depressed inpatients) and findings may not generalize to different patient groups.

Their purpose is generally to allow one to assess change, usually improvement. It is therefore important that they should be reliable. Test–retest reliability is, however, difficult to assess in practice, as patients either get used to answering the test in a particular fashion (the 'halo' effect) or their symptoms change. Computerized statistical analyses to allow for these factors are available, but are rarely employed. Scales should furthermore be reliable when used by different raters – inter-rater reliability. This is easier to assess and may be done in a variety of ways.

For these reasons most confidence is placed in established scales with extensive background data available. However, improvements are suggested often to facilitate administration, e.g. by patient self-rating, but also to improve their value, for instance by being more sensitive to change. It is important that such scales should receive general validation to ensure that they are not improving their apparent sensitivity at the expense of validity, for instance that response to treatment by a particular group of drugs reflects true improvement, and not solely the non-therapeutic changes induced by drug treatment.

1.2   The problems of investigating clinical biochemical changes

The problems of investigating biochemical changes in the brain during psychiatric disease are formidable and can most simply be stated as follows:

Where to measure?
What to measure?
What do the results obtained mean?

These will be dealt with in turn but are clearly inter-related to some extent.

Where to measure?   In general it is impossible to measure biochemical changes occurring in the brain. The only exceptions occur during neurosurgery, an uncommon event during psychiatric illness, or by analysis of post-mortem brain.

Studies on post-mortem brain have been valuable, particularly in Alzheimer's disease (Chapter 6), Huntington's chorea (Chapter 7), Parkinson's disease (Chapter 8), and to a lesser extent schizophrenia (Chapter 5). Nevertheless there are still major problems. Many neurotransmitter concentrations and enzyme activities change rapidly after death. Immediate freezing of tissue after death is invariably impossible for social, ethical, legal, and practical reasons. The cause of death can influence results; for example, oxygen deficit altering enzyme activity, because the subject died from bronchopneumonia. Finally patients in those countries where most research is carried out will almost certainly have been on drug therapy. Most drugs will influence and perhaps 'mask' pathological biochemical changes. This applies not only to transmitters and enzymes but also to receptor systems.

Clearly we would like, where possible, to have relatively non-invasive methods of study, whereby the patient can be studied when ill and, at least in those psychiatric conditions such as acute schizophrenia, depression, and mania, following remission of symptoms.

Cerebrospinal fluid concentrations of transmitters and metabolites have been studied, but the procedure of lumbar puncture is not without risk. There is also now the question as to whether CSF neurotransmitter metabolite concentration always reflects metabolic changes in the brain. Even if they do there is little evidence to suggest that metabolism and function are always closely related.

Blood and urine are usually simple to collect, although even this can be a problem, especially in very ill patients and outpatients. However, the general feeling now is that little data of value can be gained by examining transmitter precursors and metabolites in either fluid because of the small fraction of total product that is either required or alternatively produced by the brain.

Following the introduction of ligand–receptor binding techniques for studying receptor populations (Section 2.14) there has been effort expended to see whether there are changes in peripheral receptors which may reflect changes in receptor function in the brain. Some useful data on α-adrenoceptors and 5-HT receptors on platelets are now appearing, but the major assumption has to be made that any change reflects a central change.

What to measure?　The obvious answer to this is that one measures what one has the technology to measure! However, the clear corollary of this does not always seem to be realized. One cannot measure compounds for which we do not have the methodology or those compounds which have not been identified (many probably exist). If we estimate the total number of synapses in the brain and calculate the concentration of the known transmitters necessary for these terminals, we can account for only a small proportion (say 10–20%).

We measure what we can, but there is no technology available that allows one to 'explore' for new compounds. In practice what is done is to make an educated guess that compound 'X' may be in the brain, develop the methods to identify it and then extract brain tissue to see if it is present. This has happened with the peptides (Chapter 2, Part 5) and accounts for the explosive growth in the number now known to be present in the brain.

In general the compounds measured have been the neurotransmitters and their metabolites in brain or peripheral tissues; their synthetic and degradative enzymes in the brain and occasionally peripherally (such as platelet monoamine oxidase activity). Such measurements will indicate major metabolic changes, but do not necessarily indicate dynamic changes (i.e. synthesis rates) or functional changes which may not be closely linked to metabolic changes. After all, presumably what we are interested in is the function of the 'system', how much transmitter is available to stimulate the receptor, and whether on stimulation there is a change in response.

In an attempt to answer this last point various studies have now employed ligand–receptor binding techniques to examine whether receptors have been altered. Again this has been performed mainly on post-mortem brain tissue, but also occasionally on platelets and leucocytes, now it is known that they have 5-HT, dopamine, or α-receptors on their membranes. Assumptions then have to be made as to whether any peripherally measured change reflects a central change.

Other recent attempts to examine receptor function have included the use of 'challenge' tests, particularly neuroendocrine tests. Growth hormone release is under dopaminergic and adrenergic control; prolactin release inhibitory factor is probably dopamine. The technique therefore is to challenge these systems with agonists or antagonists to see if the response is abnormal in pathological states, the reasoning being that in illness there might be a general up- or down-regulation of the receptors, not only in various brain areas, but in the hypothalmic–pituitary systems as well. Such approaches are still in their infancy and hampered by the complexity of the mechanisms involved in controlling neuroendocrine release; mechanisms which are still being elucidated.

What does it mean? Lastly we come to interpretation of data. One must make sure that any differences observed have not resulted from some of the problems outlined above – drug treatment, age or sex of patient (proper control selection is always vital), storage of tissue, etc. What will still not be known is whether the observed change has any primary relationship to the disease process or whether it is secondary – that is whether the change caused the disease or whether it has occurred because of alterations in other neurotransmitter systems and has little bearing on the aetiology or pathology of the disease process. To try to answer that is a formidable problem and we often therefore use pharmacological data as a further clue.

1.3 The problems of investigating the pharmacology of psychoactive drugs

In some ways the problems of investigating the pharmacology of psychiatric illness are those associated with examining the clinical biochemical changes 'where to measure?'; 'what to measure?; and what does it mean?'. The main advantage of pharmacological research is that we can use experimental animals, but of course, that gives rise to its own problems.

If animals are given psychoactive drugs, they can be killed and changes looked for in all tissues including brain. Nevertheless there are problems in deciding the optimum dose (most doses in animals are not comparable on a dose/weight ratio to those in humans) and there can be marked species differences. Furthermore it is often important to give the drugs chronically. Both tricyclic antidepressants and ECT work to alleviate depression only when given over a period of time. Much of the earlier work on the changes produced by single doses of tricyclic or a single electroconvulsive shock has told us little about their therapeutic mechanism of action and may have been frankly misleading.

Furthermore we do not believe there is any totally satisfactory animal model of psychiatric disorder. Drugs cannot therefore be given to a 'pathological' brain to make it normal but rather to a 'normal' brain, thereby perhaps changing it to an 'abnormal' state. Changes that occur may thus be of a reflection of brain neurotransmitter systems attempting to minimize the changes being produced by the drug.

In patients, the brain is normally inaccessible. However, it is not unreasonable to suppose that if a drug alters neurotransmitter metabolism in the brain it also does so in the periphery (where appropriate). It is also reasonable to suppose that any drug acting at central receptors will also act on peripheral ones. Thus there is sense in examining peripheral neurotransmitter metabolism and receptor function following drugs and suggesting that any changes observed may well also be occurring in the brain. Drugs after all do not seek the 'target' organ but, in general, distribute themselves through the whole organism. If for example we have evidence that a monoamine oxidase inhibitor enters the brain well, then it is a reasonable assumption that a high degree of inhibition of this enzyme in the blood platelet reflects marked enzyme inhibition in the brain. Similarly a drug which is an antagonist at the 5-HT receptor on the platelet may well be acting at central receptors.

Whilst this approach looks useful it must be remembered that there are in fact very few peripheral systems that can be easily examined and thus the applications of these techniques are limited.

Finally we return to the problem of interpreting data obtained. There is a temptation (often succumbed to) to make unjustified inferences. Viz.:

> 'Drug A is useful in treating a particular psychiatric illness. Drug A decreases the concentration of transmitter X in rat brain. Therefore this drug probably works clinically by decreasing the concentration of transmitter X in the brain. Therefore there may be an overactivity in the function of transmitter X in the brains of persons suffering from this particular illness.'

The weakness of such arguments is perhaps clear but bears reiterating.

It could be that the changes seen in transmitter X have no therapeutic significance at all, but account for the side effects of drug A. Furthermore even if we are altering the mood state by altering transmitter X, the pathology of the illness need not primarily involve this transmitter. Indeed it could be that the transmitter function is normal in the illness but one can alter the mood state by altering this neurotransmitter system.

What of course should be done is to combine the experimental data obtained in animals on the biochemical changes produced by drugs used clinically, add to this our knowledge of clinical biochemical data, and hope that there emerges a pattern which will suggest what biochemical pathological changes may be occurring and how drugs may be altering this. In turn this allows further experimentation designed to test the hypothesis that has been developed.

To date there are very few data to suggest that the drugs used successfully to treat the major psychiatric disorders are acting to normalize any primary (that is causative) biochemical abnormality.

1.4 The value of psychopharmacological research

Having now read all the problems associated with experimental and clinical psychopharmacology the reader may well be wondering whether it is worth doing any research at all! Many of the drugs in current use have not been introduced into clinical use because they were 'designed' for the job: chlorpromazine was not developed as an antischizophrenic, nor mianserin as an antidepressant. Nevertheless examination of drug action on neurotransmitters does give clues as to how useful drugs may be developed. We may not know why dopamine antagonism is an important feature of antischizophrenic drugs, but it provides a useful screen in developing new compounds.

Furthermore there have been some spectacular successes. The use of L-dopa in Parkinson's disease (Chapter 8) is a prime example. Following observations on neurotransmitter concentrations in post-mortem brains a specific treatment was proposed and found to be successful. This chapter has been included in this book not because it is a psychiatric disease (although it does sometimes present with psychiatric complications) but as an example of a psychopharmacology success story!

Research on opiates has led to the discovery of the enkephalins (Section 2.29) and should lead to better methods of analgesia and greater knowledge of the problems of addiction.

The research on benzodiazepines (Chapter 4) may lead to the discovery of a natural benzodiazepine-like compound in the brain and thus to better treatment of anxiety and epilepsy.

Psychopharmacology is an area of biological research that is in its late infancy. Most of the 'established' neurotransmitters (5-hydroxytryptamine, dopamine, noradrenaline, GABA for example) were detected and mapped in the brain during the middle 1950s to late 1960s. Research on the action of drugs on the metabolism of these compounds blossomed during the 1960s and studies on receptor changes have occurred during the last 5 years. During the last 2–3 years there has been an explosive growth in our knowledge of the existence and distribution of peptides. However, we have very little information on the function of these compounds.

Information on the action of drugs used in psychiatry is now reaching a period of consolidation where earlier data is being re-evaluated and new data being obtained which are not always clearly consistent with existing hypotheses. The speed with which new areas can develop (viz. the work on enkephalins) demonstrates how exciting psychopharmacology research is at present and this excitement will doubtless continue as new hypotheses are proposed, evaluated, confirmed, or denied.

Chapter 2

Biochemistry of the Neurotransmitters

PART 1   5-HYDROXYTRYPTAMINE

2.1   Isolation and identification

The initial identification of 5-hydroxytryptamine (5-HT or serotonin) as a compound present in animal tissue was achieved independently in two countries. Erspamer in Italy spent many years characterizing the substance present in enterochromaffin cells of the intestinal mucosa, and named the isolated compound enteramine. In the United States of America Page and his colleagues had been trying to identify the potent vasoconstrictor compound present in clotted blood, which they named serotonin. In the early 1950s both compounds were identified as 5-hydroxytryptamine. Shortly afterwards both Gaddum and colleagues in Edinburgh and Page demonstrated the presence of 5-HT in mammalian brain.

Within a further year Woolley and Shaw (1954) had made the heuristic proposition that since lysergic acid diethylamide produced marked disturbances of thought and vision and had marked structural similarities to 5-HT (see Fig. 3.1) it was possible that the compound 5-HT might subserve a transmitter role in the brain and be involved in the aetiology of mental disorders. As will be seen later nothing that has been found in the subsequent 25 years has strongly discounted this proposition.

2.2   Distribution

Fluorescence histochemistry produced less detailed results on 5-HT distribution than on the catecholamines, because of lower sensitivity of the reaction. However, recent studies using techniques such as immunohistochemistry have led to good mapping. The main cell body areas and projections are shown in Fig. 2.1. The raphé region is the site of all the cell bodies and can be lesioned with neurotoxins such as 5,7-dihydrotryptamine, to produce profound decreases in 5-HT concentrations in the terminal regions. The B1–B3 groups are the descending pathways from the ventral raphé, projecting to the medulla and spinal cord. The B7–B9 cell body groups provide extensive innervation of the striatum, mesolimbic forebrain, cortex, hippocampus, thalamus, and hypothalamus.

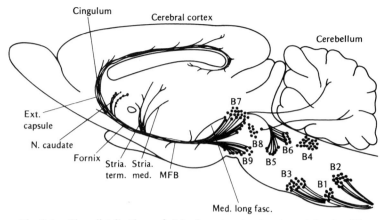

Fig. 2.1 The distribution of 5-hydroxytryptamine in rat brain. The main region of the cell bodies is the raphé. (Reproduced from Cooper, Bloom and Roth (1978) with permission of the Oxford University Press)

2.3 Synthesis

5-Hydroxytryptamine does not pass into the brain from the periphery, and therefore has to be synthesized from its amino-acid precursor L-tryptophan. The appropriate metabolic machinery for synthesis and degradation are present at the nerve ending (Fig. 2.2).

L-Tryptophan has a property unique among the amino acids in the blood in that it is bound to plasma albumin. As a result only about 15% of the total tryptophan

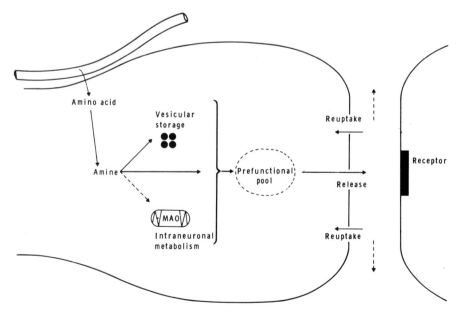

Fig. 2.2 A simplified diagrammatic representation of the nerve ending

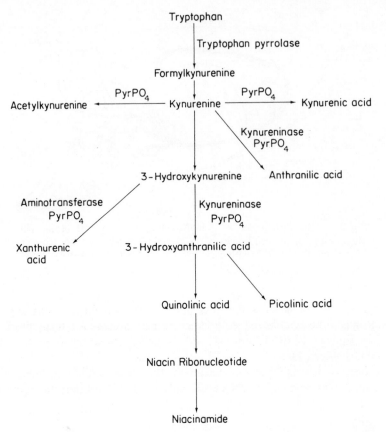

Fig. 2.3 Simplified metabolic pathway of tryptophan metabolism (kynurenine pathway). Pyr PO₄, pyridoxal phosphate

present in the plasma is 'free' or non-albumin bound. This was first reported in 1958 but it was not until 15 years later that the possible neurochemical importance was realized (see Green, 1978). There has subsequently been a great deal of controversy regarding the relative importance of 'free' versus 'total' tryptophan as a determinant of brain tryptophan concentration. However, it is now clear, mainly from the work of Curzon's laboratory in London, that free tryptophan is the fraction available for transport into the brain and alterations in the free tryptophan concentration influence brain tryptophan concentration.

At least part of the peripheral control of free tryptophan concentration resides in the activity of the enzyme tryptophan pyrrolase in the liver. Quantitatively most tryptophan is metabolized by this enzyme to kynurenine and then to a variety of other metabolic products (Fig. 2.3). The quantitatively minor pathway goes to 5-HT (Fig. 2.4). Tryptophan pyrrolase is an inducable enzyme, activity being increased both by its substrate tryptophan and by corticosteroids. Thus an increased plasma tryptophan concentration results in enzyme induction and

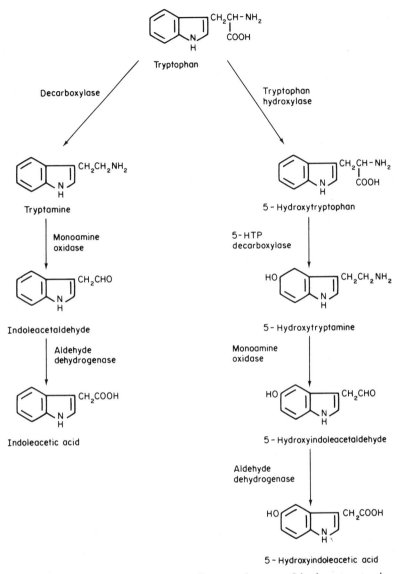

Fig. 2.4 The metabolic pathway of tryptophan to 5-hydroxytryptamine, tryptamine, and their acid metabolites

therefore a more rapid metabolism of the amino acid. It has been demonstrated that acute administration of hydrocortisone increases pyrrolase activity in rats thereby lowering bound and free plasma tryptophan, and central tryptophan concentrations, and consequently the concentration of 5-HT and 5-HIAA in the brain are also lowered. It seems reasonable to suppose that longer term or diurnal regulation of brain 5-HT synthesis might in part be accomplished by alterations in the activity of tryptophan pyrrolase, which has a diurnal variation in its activity.

The uptake system transporting tryptophan into both the brain and the nerve ending is not specific to tryptophan; other amino acids share this process and thus compete with tryptophan for uptake. These amino acids are the large neutral amino acids — leucine, isoleucine, tyrosine, valine, and phenylalanine. It seems probable that both the ratio of tryptophan to the large neutral amino acids, and the concentration of 'free' tryptophan help to determine the amount of tryptophan entering the brain (see review by Green, 1978).

Following entry into the brain tryptophan is converted to 5-HT by two enzymatic steps, hydroxylation followed by decarboxylation (Fig. 2.4).

The first enzyme on this pathway is tryptophan-5-hydroxylase. It was first demonstrated directly in the brain in 1964 and the delay was due mainly to various difficulties in assaying its activity. The enzyme has low activity, non-enzymatic hydroxylation can occur and there are difficulties in separating the product 5-hydroxytryptophan.

The distribution of the enzyme has been studied and in general the brain stem, hypothalamus, caudate nucleus, and amygdala have higher activity than other regions of the brain. The pineal and pituitary also display hydroxylating activity.

Careful homogenization of brain tissue in isotonic sucrose results in the nerve endings shearing off intact and then resealing to form pinched off nerve endings, which are called synaptosomes, and with appropriate centrifugation these can be harvested. It appears that the greater part of the activity of tryptophan hydroxylase is associated with synaptosomes and that most of the enzyme activity is associated with the soluble fraction.

The enzyme has a requirement for molecular oxygen (increasing the oxygen content of inspired air has been said to increase the rate of 5-HT synthesis) and it

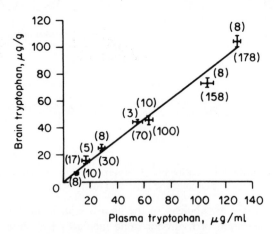

Fig. 2.5 The relationship between the concentration of tryptophan (figures in brackets give the dose of tryptophan administration in mg/kg and number of observations made) in plasma and brain of rats. (From Grahame-Smith (1971), reproduced with permission of Pergamon Press Ltd)

requires a pterine cofactor. There are problems regarding the identity of the pterine cofactor as it has been shown that addition of several pterine cofactors markedly alters the enzyme activity and it is still uncertain which is the natural cofactor.

What does seem clear is that hydroxylation is the rate-limiting step in the biosynthesis of 5-HT, the activity of the next enzyme (decarboxylase) being 80–100 times higher in some brain regions. Furthermore it also seems clear that the hydroxylase enzyme is not normally saturated with its substrate. The result of this is that increasing the amount of tryptophan in the brain (by administration of tryptophan) increases the rate of 5-HT synthesis (Grahame-Smith, 1971; Figs. 2.5 and 2.6).

In neuropharmacology several compounds have been used as hydroxylase inhibitors, the most extensively used being *p*-chlorophenylalanine (PCPA). *In vivo* PCPA appears to act as an irreversible inhibitor and following administration of the compound in high dose (200–300 mg/kg) the brain 5-HT concentration decreases by 60–70% and remains low for around 2 weeks. Another inhibitor used with great success is *p*-chloramphetamine, which also releases 5-HT from nerve endings, blocks reuptake, and acts as a monoamine oxidase inhibitor. The immediate effects of its administration is the appearance of 5-HT mediated behavioural excitation (see for example Green and Kelly, 1976). This drug also has the interesting long term effect of being neurotoxic to the B6 group of neurons and therefore unlike 5,6-dihydroxytryptamine or 5,7-dihydroxytryptamine, which do not pass the blood–brain barrier, is able to destroy some central 5-HT neurons

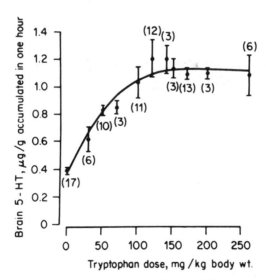

Fig. 2.6   The accumulation of brain 5-HT during 1 hour following administration of different doses of L-tryptophan. (From Grahame-Smith (1971), reproduced with permission of Pergamon Press Ltd)

14

following its peripheral injection. These data clearly preclude the clinical use of *p*-chloramphetamine; PCPA on the other hand has been used as a research drug in several investigations of human 5-HT metabolism (for example in patients with the carcinoid syndrome).

The decarboxylase enzyme was first identified in 1954 in the guinea pig kidney. Following improved purification techniques and studies of the distribution of the enzymes decarboxylating 5-hydroxytryptophan (5-HTP) and L-dihydroxy-phenylalanine (L-dopa) it was concluded that these were the same enzyme and it was therefore named aromatic L-amino-acid decarboxylase. Recent data have suggested that there may be two enzymes, with the observations that the pH and temperature optima for the two reactions are different and that brain 5-HTP decarboxylase activity is not reduced following destruction of central dopamine terminals. It has also been suggested that there is no parallel distribution of the two enzymes (see review of Green and Grahame-Smith, 1975). Nevertheless even if there are two distinct decarboxylase enzymes for 5-HTP and L-dopa it is clear that there is little substrate specificity and the appearance of 5-HT in dopamine terminals after 5-HTP loading indicates 'non-specific decarboxylation' of 5-HTP. This does not happen after L-tryptophan loading since the tryptophan hydroxylase enzyme is quite different from tyrosine hydroxylase and is located only in 5-HT-containing neurons.

The 5-HTP decarboxylase enzyme has been identified in synaptosomes and appears to be about 60% particulate bound although much is clearly in the cytoplasm. It is an enzyme which requires pyridoxal phosphate as a cofactor.

The best known inhibitor of the decarboxylase enzyme is α-methyldopa (this drug is used in the treatment of hypertension, but is not thought to work by inhibiting decarboxylase). The high decarboxylase activity in the brain, however, results in even high and repeated doses of α-methyldopa and other inhibitors (see Green and Grahame-Smith, 1975) having relatively little effect on brain 5-HT concentration.

Recent interest has centred on peripheral decarboxylase inhibitors and the fact that α-methyldopahydrazine administration allows a lowering of the dose of L-dopa without a reduction in the therapeutic effect of this drug in Parkinson's disease, attests to the effectiveness of this approach (Section 8.13).

2.4 Metabolism

Degradation of 5-HT to its major metabolite 5-hydroxyindole acetic acid (5-HIAA) is also achieved by a two step process, oxidation followed by dehydrogenation (Fig. 2.4).

The first enzyme on the degradative pathway is monoamine oxidase. There has been enormous interest in this enzyme, and its inhibition, because of the therapeutic use of the inhibitors as antidepressant drugs. For a comprehensive review of this enzyme there are several recent books. In particular those of Costa and Sandler (1972) and Wolstenholme and Knight (1976) are to be recommended.

In brain the enzyme is almost completely associated with the mitochondrial function and is probably located on the outer mitochondrial membrane. It is an enzyme having limited substrate specificity and is able to deaminate many amines having the formula $R-CH_2-NH_2$ with R being a substituted aryl or alkyl group. The physiological substrates are the biogenic amines, 5-HT, tyramine, noradrenaline (norepinephrine), adrenaline (epinephrine), and dopamine.

Over the last few years there has been interest in the possible existence of multiple forms of this enzyme. In particular attention has focused on the possible division of the enzymes into two forms – type A and type B. This concept arose following the observations that the inhibitors clorgyline and deprenil failed to show the kinetics expected from a single enzyme system, but that consistent with there being two enzyme systems (for example see Johnston, 1968). It was suggested that type A is inhibited by clorgyline, and its substrates are noradrenaline and 5-HT while type B which is inhibited by deprenil and metabolizes benzylamine and β-phenylethylamine (Table 2.1). Neff and Goridis (1972) showed differential distribution of the two forms in different rat tissue (e.g. 15% type A and 85% type B in pineal), while rat brain contains approximately equal amounts of type A and type B and at the time it seemed reasonable to suggest that the ratio of the forms was indicative of physiological requirements. However, subsequent experiments have not encouraged this view. In human brain for example it has been claimed that the MAO is almost exclusively type B (Glover et al., 1977) and this suggests that the pharmacological typing does not reflect the physiological activity. The fact that the physiological substrates do not always fall into the same groupings when different inhibitors are used also bears out this view. Furthermore it now seems probable that type A and type B refer not to absolute substrate requirements but rather to relative affinities of the substrates for the two enzyme forms. For example both in the brain and the liver complete inhibition (as measured in vitro) of type A MAO does not totally inhibit 5-HT degradation (in vivo) and it is only when type B is also inhibited that 5-HT

Table 2.1 Substrates and inhibitors for type A and type B MAO

	Type A MAO	Type B MAO
Preferred substrate	Noradrenaline 5-Hydroxytryptamine	β-Phenylethylamine Benzylamine
Substrates for both types		Dopamine Tyramine Tryptamine
Non-specific inhibitors		Pargyline* Tranylcypromine Iproniazid
Specific inhibitors	Clorgyline Lilly 56141	Deprenil

*Pargyline is a type A inhibitor at low dose

metabolism is blocked. This demonstrates that when type A MAO is inhibited (with clorgyline) type B MAO continues to metabolize the amine to some extent.

The conventional MAO inhibitors, pargyline, tranylcypromine, phenelzine and iproniazid (Table 2.1) are all non-specific inhibitors and inhibit both forms of the enzyme. In rat brain over 85% inhibition of MAO must be achieved before there is a large increase in either 5-HT content or function (Green and Youdim, 1975) and it seems probable that such high values may not often be reached during clinical administration of MAO inhibitors. A study on the post-mortem brain tissue of patients dying whilst on MAO inhibitors found that inhibition of MAO by most inhibitors was in the region of 60–70% (Youdim *et al.*, 1972).

Further discussion of MAO in regard to its function in catecholamine metabolism is found in Section 2.9 and the clinical use of MAO inhibitors in the chapter on their use in depression (Chapter 3). That section also deals with the problems of the food-induced interactions.

Oxidation of 5-HT by MAO produces 5-hydroxyindoleacetaldehyde which can either be dehydrogenated by aldehyde dehydrogenase to 5-HIAA or reduced by aldehyde reductase to the alcohol 5-hydroxytryptophol (5-HTOH). It has been shown that 5-HTOH can be detected in human urine and both products occur in the liver. Furthermore the metabolism can be shifted from the acid to the alcohol by ethanol administration, but it seems that this probably does not occur in the brain, and 5-HTOH may not even be formed in the brain (see review by Green and Grahame-Smith, 1975). 5-Hydroxyindoleacetaldehyde cannot normally be detected in the brain, except by use of trapping agents, which complex with the aldehyde irreversibly, suggesting that the aldehyde dehydrogenase step must be very rapid.

2.5 Control of synthesis, metabolism, and function

Following release, monoamines (including 5-HT) are inactivated by a high affinity, energy requiring uptake process into the nerve ending (Iversen, 1975b) and tricyclic anti-depressants inhibit this (Section 3.25). The question arises, however, as to what mechanisms control the synthesis, release and subsequent metabolism of 5-HT following re-uptake and whether regulation of 5-HT synthesis controls 5-HT function.

First it is unlikely that alterations of MAO activity play a major role in controlling 5-HT metabolism. It has been pointed out that the enzyme is present in 'excess' insofar as increasing 5-HT function requires a high degree of inhibition of the enzyme. Also the half-life of the enzyme in the brain is between 4 and 9 days, ruling out rapid changes in the amount of enzyme present. Presumably the MAO which plays a vital role in controlling 5-HT metabolism is that present within the nerve ending and this is a very small proportion of the total MAO present in the brain. Destruction of dopamine terminals by administration of the neurotoxin 6-hydroxytryptamine does not result in a measurable lowering of brain MAO activity, perhaps an expected finding in view of our knowledge that this enzyme is present in mitochondria and thus has an ubiquitous distribution. Until we can preferentially inhibit intraneuronal MAO it is difficult to assess whether the enzyme can play an active role in controlling the metabolism of monoamines. Although it clearly plays a passive role.

As discussed earlier, tryptophan availability certainly alters the rate of 5-HT synthesis; a change in peripheral tryptophan availability altering central 5-HT metabolism. However, availability is controlled by a variety of mechanisms and is thus a complex matter. The ratio of free to bound tryptophan in plasma (which is influenced by the non-esterified fatty acid concentration and also by various drugs); the ratio of tryptophan to other large neutral amino acids competing for uptake both into the brain and into the nerve endings; the effect of insulin in altering the uptake of amino acids into tissues; and the activity of hepatic pyrrolase, all influence the rate of brain 5-HT synthesis. Since several of these factors show regular cyclic variations (pyrrolase activity and food intake for example) it seems reasonable to suggest that the longer term diurnal regulation of 5-HT function is controlled by these changes in tryptophan availability.

What does seem unlikely is that changes in tryptophan availability and 5-HT synthesis are directly and intimately involved with the regulation of 5-HT function (function for this purpose being defined as the transmitter which has been released, has stimulated the postsynaptic receptor, and has produced some physiological response). In order to accept the hypothesis that synthesis and function are intimately linked we have to accept that synthesis is geared for moment to moment changes in release. It seems most unlikely that the many factors controlling synthesis could be interlinked and controlled to provide such a fine control. Indeed Curzon, Fernando and Marsden (1978) have demonstrated that synthesis does not even necessarily change following neuronal stimulation, thereby demonstrating clearly that the mechanisms responsible for synthesis are not immediately linked to those responsible for release.

What has been proposed elsewhere (Grahame-Smith, 1971; Green and Grahame-Smith, 1975) is that 5-HT is made in excess of functional needs and a proportion of transmitter synthesized is metabolized intraneuronally by MAO without having been released and becoming functionally active. It has been shown that L-tryptophan administration to rats markedly increases 5-HT synthesis without producing any overt behavioural changes. However, where MAO is inhibited marked 5-HT induced behavioural changes occur (Grahame-Smith, 1971). This demonstrates that increased 5-HT synthesis need not cause increased transmitter release unless MAO is also inhibited. Similar data have been obtained recently by Marsden et al. (1979) who used an electrochemical detector implanted in rat brain, and were unable to detect increased transmitter release following L-tryptophan, unless MAO had also been inhibited.

It appears therefore that changes in compartmentation of transmitter at the nerve ending (that is transmitter which may be earmarked for release or for metabolism by MAO) may play a vital role in controlling 5-HT function. These compartments are probably in equilibrium and an alteration in function can thereby be quickly achieved by altering the rate of intraneuronal metabolism.

If we acknowledge that rates of synthesis (or basal concentrations) do not reflect function this has important implications for much of the data obtained on transmitter synthesis in clinical studies since it means that a decreased (or increased) rate of synthesis does not indicate altered function, though of course it does suggest a changed rate of metabolism. This distinction does not always seem to have been appreciated.

PART 2   THE CATECHOLAMINES

2.6.   Isolation and identification

Under the general term catecholamines are grouped three neurotransmitters: dopamine (DA), noradrenaline (NA, norepinephrine) and adrenaline (epinephrine).

Although 5-hydroxytryptamine does occur peripherally (platelets, gastro-intestinal system), in neuronal tissue its distribution is exclusive to the central nervous system and the pineal. In contrast noradrenaline is distributed in the peripheral adrenergic nervous system, as was first shown by von Euler in 1946. Subsequently it was demonstrated that this compound was also distributed in a non-uniform manner in the brain.

It soon became clear that dopamine was not present in the brain merely as a NA precursor, since its distribution pattern is quite different (next section).

2.7   Distribution

The distribution of dopamine is rather limited and (if we ignore the rather specific and localized neuron systems) may be conveniently divided into three major systems (Fig. 2.7).

The first system is the nigrostriatal pathway, with cell bodies in the substantia nigra, mainly the zona compacta (A9 region) and projecting into the striatum, containing the nucleus caudatus (caudate nucleus) and putamen.

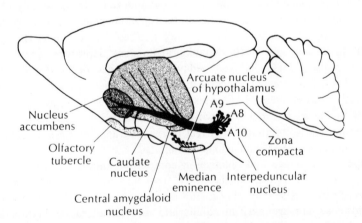

Fig. 2.7   The distribution of dopamine in rat brain. (Reproduced from Cooper, Bloom and Roth (1974) with permission of the Oxford University Press)

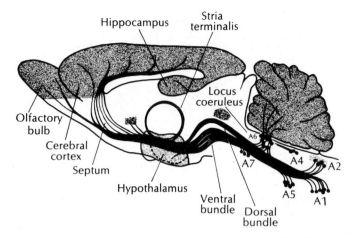

Fig. 2.8 The distribution of noradrenaline in rat brain. Stippled areas indicate areas of highest innervation. (Reproduced from Cooper, Bloom and Roth (1974) with permission of the Oxford University Press)

The second system has cell bodies in the interpeduncular nucleus (A10) and project on a tract lying close to the nigrostriatal tract to the mesolimbic forebrain (nucleus accumbens and olfactory tubercles).

The third pathway is that of the tuberoinfundibular system with the cells mainly in the arcuate nucleus and projecting to the median eminence.

Noradrenaline cell bodies are present in the pons medullary region (Fig. 2.8). The locus coeruleus has received much attention since lesioning of this region can produce marked decreases in NA concentration in cortex and hippocampus. Noradrenergic projections to the forebrain pass along the ventral and dorsal bundle and noradrenaline has a wide distribution including the forebrain hypothalamus and cerebellum.

Studies on the distribution of adrenaline in the central nervous system are recent. There are two major cell body groups, in the lateral tegmental system and dorsal medulla, with projections to the hypothalamus, locus coeruleus and spinal cord.

2.8 Synthesis

The catecholamines are formed in the brain from their amino-acid precursor tyrosine (Fig. 2.9). Phenylalanine is converted to tyrosine in the liver, by the action of phenylalanine hydroxylase. The enzyme is not present in the brain so strictly speaking, phenylalanine is not a catecholamine precursor although it could be said to be an indirect precursor.

Tyrosine is transported into the brain and into the nerve endings by the same uptake process as is employed by tryptophan. However, unlike the effect of tryptophan concentration on 5-HT synthesis, changes in tyrosine concentration

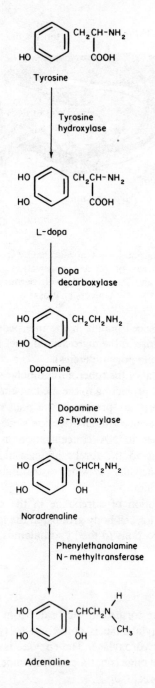

Fig. 2.9 The synthesis of catecholamines from their amino-acid precursor L-tyrosine

do not have dramatic effects on catecholamine synthesis. Indeed, tyrosine loading has little effect on catecholamine synthesis.

Tyrosine hydroxylase, an enzyme shown to be present in all catecholamine containing tissues (e.g. sympathetic nerves, adrenella medulla) produces dopa from tyrosine. It is an enzyme of high substrate specificity (it will not hydroxylate tryptophan and therefore 5-HT is not formed in catecholamine neurons). It has cofactor requirements of oxygen, Fe^{2+} and a pteridine. In the brain the enzyme is present in the synaptosomal fraction.

Hydroxylation appears to be the rate-limiting step in catecholamine biosynthesis. Inhibition of the enzyme can totally suppress catecholamine synthesis and thus decrease DA and NA concentrations in the brain. Various compounds have been used as tyrosine hydroxylase inhibitors but the main drug used in experimental psychopharmacology and which has been used in some clinical investigations is α-methyl-p-tyrosine.

The second metabolic step is decarboxylation by dopa decarboxylase, producing dopamine. Like 5-HTP decarboxylase, this enzyme is very active and whilst it may be the same enzyme as 5-HTP-DC, there is some evidence to suggest that it is different, but closely related (see Section 2.4). It has been proposed that there is a single enzyme protein with two distinct substrate sites and a single catalytic site. L-Dopa is very difficult to detect in brain, presumably because of the high activity of the decarboxylase enzyme.

In those regions in which dopamine is the neurotransmitter, these two steps, hydroxylation and decarboxylation, represent the two synthetic steps for dopamine and further metabolic steps are degradative. In noradrenergic nerves, on the other hand, dopamine is merely the immediate precursor of the neurotransmitter noradrenaline. The enzyme involved in this step is dopamine-β-hydroxylase, an enzyme requiring molecular oxygen and ascorbate. It is a copper containing enzyme and this fact is utilized in inhibitor studies. It appears to be a particulate enzyme, associated with amine storage vesicles. This localization explains in part why stimulation of peripheral sympathetic nerves results not only in release of noradrenaline but also dopamine β-hydroxylase.

Dopamine β-hydroxylase is an enzyme with limited substrate specificity. Administration of α-methyldopa thus results in the formations of α-methyl dopamine and thence to α-methylnoradrenaline, a compound that replaces noradrenaline at the nerve ending.

The enzyme can be inhibited by various copper chelating agents such as diethyldithiocarbamate, FLA-63, and disulfiram (a drug used to treat alcohol addiction – Section 9.5). Disulfiram will decrease central noradrenaline stores markedly whilst sparing dopamine. Administration of dihydroxyphenylserine (DOPS) will replace noradrenaline stores in animals with depleted noradrenaline concentrations following α-methyl-p-tyrosine administration as it is β-hydroxylated to noradrenaline.

Finally it has been shown in adrenal medulla that noradrenaline is N-methylated by phenylethanolamine-N-methyltransferase to form adrenaline. This enzyme has also been demonstrated in the brain. Like other N-methylating enzymes it uses as the methyl donor S-adenosyl methionine.

2.9 Metabolism

The metabolism, and thus inactivation, of both dopamine and noradrenaline is achieved by similar metabolic steps (Fig. 2.10). Two enzymes are involved in the initial metabolism; monoamine oxidase and catechol-O-methyltransferase (COMT). Monoamine oxidase has been discussed in Section 2.4. This enzyme metabolizes dopamine to 3,4-dihydroxyphenylacetaldehyde and noradrenaline to

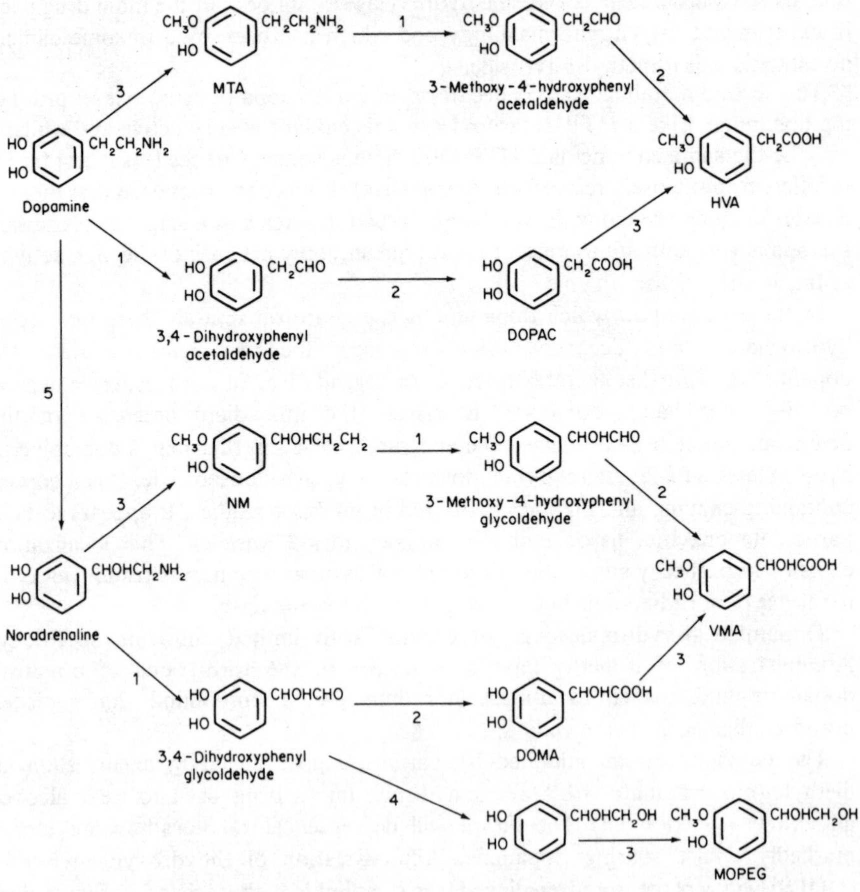

Fig. 2.10 The metabolic pathways of dopamine and noradrenaline. Enzymes are numbered as follows: 1, monoamine oxidase; 2, aldehyde dehydrogenase; 3, catechol-O-methyl transferase; 4, aldehyde reductase; 5, dopamine-β-hydroxylase. The abbreviations are as follows: MTA, 3-methoxytyramine; HVA, homovanillic acid; DOPAC, dihydroxyphenylacetic acid; NM, normetanephrine; DOMA, dihydroxymandelic acid; VMA, vanillylmandelic acid; MOPEG, 3-methoxy-4-hydroxyphenylglycol

3,4-dihydroxyphenylglycolaldehyde. These aldehydes are then dehydrogenated to the corresponding acids dihydroxyphenylacetic acid (DOPAC) and dihydroxymandelic acid (DOMA), and these are then methylated by COMT. If the catecholamines are first acted on by COMT they are methylated to 3-methoxytyramine or normetanephrine, involving the methyl donor system S-adenosyl methionine. The COMT enzyme is cytoplasmic and can be inhibited by pyrogallol and tropolones. The methylated compound is then metabolized by the MAO/dehydrogenase system. The end products are thus, in the case of dopamine, homovanillic acid (HVA) and in the case of noradrenaline, vanillylmandelic acid (3-methoxy-4-hydroxymandelic acid, VMA).

In addition, in the case of noradrenaline metabolism, 3,4-dihydroxy-phenylglycolaldehyde can be reduced by aldehyde reductase to 3,4-dihydroxy-phenylglycol (MOPEG or MHPG). This compound exists in urine in both free and as a sulphate ester conjugate. There have been several studies suggesting that in some animal species, including man, MOPEG or its sulphate ester are the main noradrenaline metabolites in the brain and it has further been suggested that most urinary MOPEG is formed from the brain with VMA being the peripheral noradrenaline metabolite.

2.10 Control of synthesis and release

The regulation of catecholamine synthesis and release is a complex area and readers wishing to learn more about this area in detail should consult some of the recent books on this area (e.g. Usdin, 1974; Iversen, Iversen and Snyder, 1975).

When a MAO inhibitor is given catecholamine concentrations rise very little. This is in marked contrast to 5-HT concentrations which rise linearly for at least 2 hours after MAO inhibition.

Tyrosine hydroxylase activity does seem to be an important controlling factor. However end product inhibition (negative feedback control; that is, inhibition by dopamine or noradrenaline of this enzyme) does not seem to be a major factor in the control process.

Short term regulation of this enzyme appears to be associated with a change in the affinity of the soluble hydroxylase enzyme for its cofactor. This apparently requires nerve impulse flow. Longer term changes involve transsynaptic activation of the enzyme by a system involving cyclic nucleotides.

However, increasing the level of activity of adrenergic neurons produces an increase, not only to tyrosine hydroxylase activity but also the activity of dopamine-β-hydroxylase and phenylethanolamine-N-methyltransferase. In the case of dopamine-β-hydroxylase this is due to formation of new enzyme protein.

It appears that newly synthesized transmitter is released preferentially.

2.11 Presynaptic receptors

In analysis of the control of catecholamine release and function, some discussion of presynaptic receptors (also called autoreceptors) is required. Until relatively recently it was not realized that specific receptors were present on the nerve ending membrane.

Much of the evidence has been obtained on peripheral nervous systems, but there are good indications that the same general phenomena occur in the central nervous system.

Presynaptic receptors monitor the concentration of transmitters within the synaptic cleft. Such a system obviously allows rapid control of the release and thus the synthesis of neurotransmitters. Monoamine agonists can be shown to inhibit synthesis, whilst dopamine antagonists increase dopamine turnover (though surprisingly 5-HT antagonists do not increase 5-HT turnover).

The increase in dopamine turnover has been convincingly shown not necessarily to require a neuronal feedback loop. When a dopamine agonist is added to a preparation of pinched off nerve endings (synaptosomes), tyrosine hydroxylase activity in the synaptosomes alters. Nevertheless *in vivo* inhibition of turnover can well occur though neuronal feedback loops of the type existing in the striatonigral pathway.

Some drugs seem either to act preferentially on the presynaptic receptors or to have dose related effects. Thus low doses of clonidine act specifically on the presynaptic α-receptor and whilst very low doses of apomorphine (say, 0.05 mg/kg) act only on the presynaptic receptor, higher doses (0.1 mg/kg) also stimulate the postsynaptic receptor. Similarly some antagonists also show specificity – yohimbine on the presynaptic α_2-receptor and prazocin on the postsynaptic α_1-receptor.

An excellent review on presynaptic receptors is that of Langer (1977). Fig. 2.11 shows his interpretation of data obtained on peripheral adrenergic nerves and indicates that both inhibitory α-receptors and positive feedback β-adrenoceptors are present, irrespective of the nature of the postsynaptic adrenergic receptor.

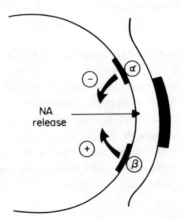

Fig. 2.11 Postulated types of presynaptic adrenoceptors, inhibitory α-receptors and facilitatory β-receptors

2.12 Turnover of neurotransmitters

Consideration of the concept of turnover must also be made in any discussion of the control of neurotransmitter release.

The term 'turnover' refers to a system of renewal of a compound in the body or in a tissue. The renewal can occur by synthesis in a given compartment or by synthesis elsewhere and subsequent transport to the compartment.

Various assumptions in measuring neurotransmitter turnover must be made:

1. the system is 'open' and transmitter is made by synthesis and lost by degradation;
2. formation and degradation rates are equal;
3. no distinction is made between newly synthesized and old molecules during degradation;
4. turnover rates of precursor and product are constant, although not necessarily equal, during measurement.

Several techniques are available.

1. Inhibition of synthesis: the synthesis is inhibited and the rate of decline of transmitter product is examined.
2. Infusion of labelled precursor: the incorporation rate of label into product is examined.
3. Pulse injection of labelled precursor: steady state is not reached but this method gives an indication of an incorporation rate.
4. Accumulation of monoamine metabolites following probenecid: the efflux of acid metabolites from the brain is prevented by this drug and the rate of accumulation of metabolite is measured.
5. Measurement of 5-hydroxytryptamine accumulation following MAO inhibition: this technique cannot be applied to catecholamines since because of feedback inhibition, there is not a linear accumulation of amine after MAO inhibition.
6. Use of ^{13}C or ^{18}O labelled precursors: this allows a measure of rate of incorporation using mass spectrometry to measure endogenous and labelled transmitter.

Clearly the non-isotopic methods are simpler and cheaper. Comparison of turnover rates using the various methods reveals broad agreement. Since the different methods rely on different assumptions, this suggests that the results have validity in indicating neurotransmitter turnover. However, there are occasions when an altered turnover rate is not related to altered functional activity of the transmitter.

Several excellent reviews of amine turnover methodology have been published (e.g. Neff and Tozer, 1968; Costa, 1970).

2.13 Dopamine- and noradrenaline-sensitive adenylate cyclase

Cyclic AMP is a mediator in the action of a number of hormones. Interactions of the hormone with the receptor enhance the activity of the membrane bound

Fig. 2.12 Proposed roles for cyclic AMP and protein phosphorylation in neuronal function include regulation of neurotransmitter synthesis in presynaptic terminals (a), regulation of microtubular function (b) (indicated, purely for convenience, as occurring in the presynaptic axon; microtubules occur in dendrites, soma and axon of neurons), and generation of postsynaptic potentials (c). The sequence of events by which neurotransmitter, released from presynaptic terminals, produces an electrophysiological response in the postsynaptic cell is conceived as follows. The released neurotransmitter activates a neurotransmitter-sensitive adenylate cyclase present in the membrane of the postsynaptic cell, leading to the production of cyclic AMP in the immediate vicinity of the postsynaptic membrane. The newly formed cyclic AMP activates a cyclic AMP dependent protein kinase present in the postsynaptic membrane. This activated protein kinase catalyses the phosphorylation of a substrate protein also present in the postsynaptic membrane, converting it from the non-phosphorylated to the phosphorylated state. A key element of this model is that this substrate protein controls the permeability of the postsynaptic membrane. Phosphorylation of the substrate protein leads, through a change either in ion conductance or in the rate of an electronic pump, to a change in membrane potential, the 'postsynaptic potential'. Since the postsynaptic potential is transient in nature, enzymatic machinery must exist which terminates this sequence of events. This enzymatic machinery includes a phosphodiesterase that hydrolyses the cyclic AMP to 5'AMP and a phosphoprotein phosphatase that converts the substrate protein back to the non-phosphorylated form, leading to the termination of the postsynaptic potential. (Reprinted by permission from *Nature*, **260**, 104. Copyright © 1976 Macmillan Journals Limited)

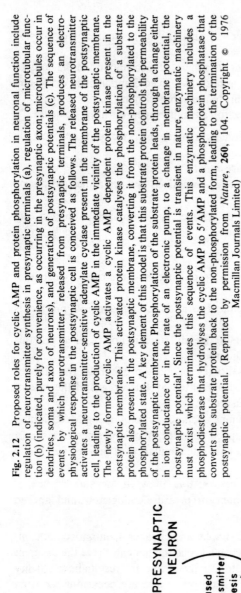

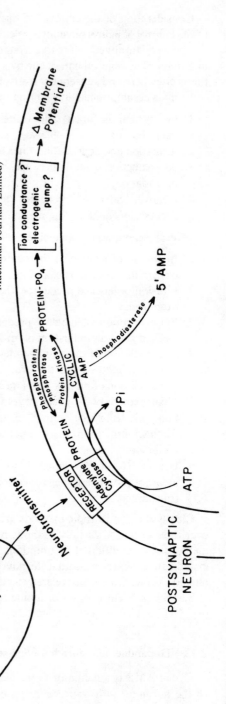

enzyme adenylate cyclase. This in turn results in an increased production of cyclic AMP from ATP. The cyclic AMP formation is the first in a series of biochemical reactions leading ultimately to the physiological response. It is this series of changes which has led to cyclic AMP being referred to as the 'second messenger'.

The brain contains relatively high concentrations of both the adenylate cyclase enzyme and the cyclic AMP catabolizing enzyme phosphodiesterase.

A noradrenaline-sensitive adenylate cyclase was reported in 1962, although activity was low in the preparation used.

In 1972 Kebabian, Petzold and Greengard demonstrated a dopamine-sensitive adenylate cyclase. A 5-HT-sensitive adenylate cyclase system has also more recently been reported (Pagel *et al.*, 1976).

A mechanism of protein phosphorylation following neurotransmitter activation has been proposed by Greengard (1976) and is shown in Fig. 2.12.

Adenylate cyclase catalyses the conversion of ATP to cyclic AMP. It requires a divalent cation (normally Mg^{2+}). At different concentrations calcium can activate or inhibit the enzyme, suggesting a regulatory role. However, whilst removal of Ca^{2+} lowers basal enzyme activity it does not alter neurotransmitter stimulation.

The enzyme is present in cell membranes. Cuatrecasas (1974) has suggested that the regulatory subunit of the enzyme (presumably the receptor) and the catalytic subunit of adenylate cyclase are separate and mobile within the membrane. Neurotransmitter activation of the regulatory subunit produces a conformational change which binds the two subunits together and activates the enzymes.

The adenylate cyclase systems certainly show characteristics of the receptor type. Dopamine-sensitive adenylate cyclase for example is inhibited by dopamine antagonists and stimulated by dopamine and dopamine agonists. β-Adrenergic antagonists have little effect, demonstrating the selectivity of the system. The possible clinical relevance of these observations is discussed further in Section 5.16.

2.14   Ligand–receptor binding techniques

The technique of ligand-receptor binding has markedly altered CNS neuropharmacology research. There is no particular reason for placing this section with the catecholamines since most neurotransmitters have now been investigated with the technique. However, in this book reference to it will be made mostly in regard to dopamine function.

The objective of ligand binding studies is to examine the interaction between a chosen ligand and a tissue receptor. A receptor is a site that will selectively bind a compound having a specific chemical structure and which, following the binding of certain compounds will produce a specific and characteristic response. Compounds acting at the receptor which generate the response are agonists whilst those that bind without producing the response, and which therefore inhibit the effect of agonists are antagonists. In ligand binding studies, a ligand, which may be an agonist or antagonist, is bound to the receptor allowing analysis of the binding characteristics (affinity of ligand for the receptor) and the number of receptors present. The technique does not allow one to differentiate between agonist and antagonist actions.

In theory the technique is simple. Tissues containing the receptor, such as brain membrane preparations, are incubated with the ligand, usually labelled with radioactivity. Following separation of the tissue from the medium the amount of ligand binding to the tissue is measured. In practice other factors have to be taken into account. The ligand will almost certainly bind to non-receptor sites in the tissues and possibly to non-biological material such as the reaction vessel wall. Binding is thus referred to as 'specific' (that is receptor) and 'non-specific' (other sites).

The technique used to separate these two components is to incubate tubes in parallel. Into both tubes is placed the radio labelled ligand; however, to one tube only, a larger concentration of another compound with high affinity for the receptor is also added. The added compound will compete with the radioligand for binding to the receptor. Since receptor binding is normally of high affinity but reversible the added drug will displace the radioligand. In contrast non-specific binding is usually of lower affinity but less reversible. The assumption is that ligand binding in the presence of the other compound thus represents non-specific binding (fig. 2.13).

Specific binding must fulfil a number of criteria.

1. Since there is a finite number of receptors within the tissue, binding must be saturable.
2. The concentrations over which the ligand binds should be comparable to those over which it is active physiologically.

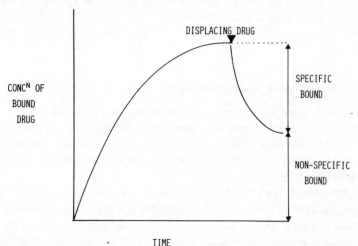

Fig. 2.13 Idealized plot of binding of radioligand to a brain membrane preparation and the displacement by a high concentration of non radioactive displacing drug, showing the specific and non-specific bound fractions

3. Specificity and stereospecificity of compounds competing for the receptor site should be similar to those seen in intact tissue.
4. The kinetics of association and dissociation of ligand and receptor should be of the order seen in the receptor response.
5. The existence of the postulated receptor in a particular tissue should be confirmed by functional responses in intact tissues.
6. Changes in ligand binding caused by drugs or disease should be reflected in changes in the functional response of the tissue.

The main method of analysis of binding data is that of Scatchard, who originally used the technique to analyse the binding of small molecules to large proteins. A range of ligand concentrations is used and measurement of the bound and free ligand at these different concentrations is made.

For a single binding site a plot of bound/free against bound results in a straight line. The intercept B gives the value of B_{max} or the maximum number of binding sites. The slope gives the value of the binding affinity. The reciprocal of this value gives the dissociation constant of K_d. Thus this analysis tells us how strongly the ligand is binding to the receptor and how many receptors are present in the tissue (Fig. 2.14).

Some tissues contain more than one receptor population, often a high affinity site of low capacity and a lower affinity site with higher capacity. The Scatchard analysis of data obtained in such tissues may clearly show these two sites (Fig. 2.14).

The ligands used for investigating the major neurotransmitters are shown in Table 2.2.

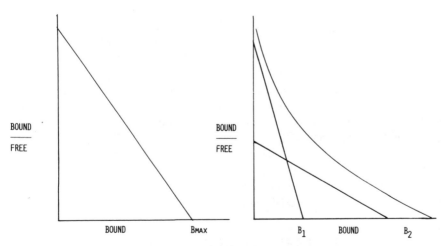

Fig. 2.14 Idealized Scatchard analyses of binding data. On the left is seen a single receptor population result with B_{max}. On the right is the plot obtained when there are two binding sites. Analysis of these data produces two straight lines, a high affinity low capacity site with B_1 as the maximum number of receptors (B_{max}) and a lower affinity higher capacity site with B_{max} at B_2. The curve crosses the axis at $B_1 + B_2$

Table 2.2 Some neurotransmitters or receptors and the radio-ligands used to examine their receptor function

Transmitter or receptor	Radioligand
5-Hydroxytryptamine	LSD 5-HT Spiperone (cortex)*
Dopamine	Apomorphine Spiperone ADTN
Noradrenaline	Dihydroalprenolol (β)† Dihydroergocriptine (α) Prazosin (α_1) WB4101 (α_1) Clonidine (α_2)
GABA	GABA Muscimol
Acetylcholine	Quinuclidinylbenzilate (muscarinic) α-Bungarotoxin (nicotinic)
Benzodiazepine	Diazepam Flunitrazepam
Opiate	Naloxone Enkephalins
Histamine	Histamine Pyrilamine (H_1) Cimetidine (H_2)

*In cortex 5-HT receptors labelled by [^{3}H]-5HT have been designated 5-HT$_1$ and those with [^{3}H]-spiperone, 5-HT$_2$ (Peroutka and Snyder, 1979).
†No very satisfactory specific β_1 and β_2 ligands are available and receptor type has been assessed by displacement with non-labelled β_1 or β_2 antagonists.

Ligand binding can therefore be seen to be a powerful method for investigating the characteristic of receptors in disease (see the changes in post-mortem schizophrenic brains; Section 5.10) and the effects of drugs on receptor function (for example, long term effects of neuroleptic drugs on dopamine receptors, Section 5.17). However, it should be remembered that it gives no indication of what takes place distal to the receptor.

The book of Yamamura, Enna and Kuhar (1978) gives further details on both the theory and practice of ligand receptor binding.

PART 3   ACETYLCHOLINE

2.15   Introduction

The role of acetylcholine as a neurotransmitter in the peripheral nervous system has been known for many years, but its role in the CNS is only now being clarified. It may be that the slowness with which our knowledge of acetylcholine in the brain has developed has been due in part to the problems in assaying the neurotransmitter.

Several assay methods are now available. The classical technique of bioassay is very sensitive but is liable to interference from other compounds. Gas chromatography is less sensitive but specific, albeit tedious. Gas chromatography–mass spectrometry (mass fragmentography) combines high sensitivity with absolute specificity, but is very expensive.

Another major difficulty in examining acetylcholine (ACh) function is that of investigating precursor–product relationships of a compound that is metabolized to its precursor.

2.16   Synthesis

The precursor of acetylcholine is choline. It is probable that choline is not formed in the brain; however, it can be synthesized in the liver, and choline and its phospholipid form, phosphatidylcholine, are transported from blood to brain. The precise role of phosphatidylcholine in ACh synthesis is not yet clear.

Choline can be transported by a low affinity transport system but also by a high affinity saturable system dependent on Na^+ and energy. This high affinity

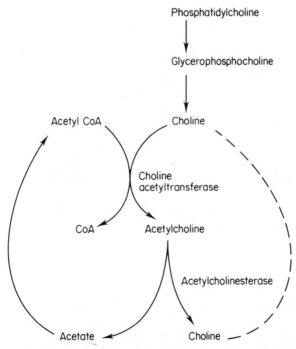

Fig. 2.15   Acetylcholine synthesis. Dotted line shows the recycling of choline following the metabolism of acetylcholine

uptake system has recently been examined in some detail since it appears to be specific for cholinergic terminals and there is some evidence to suggest that it is closely linked with ACh synthesis and release. It has been suggested that measurement of high affinity uptake of choline (HAUC) allows a measure of change in ACh function though this remains controversial.

HAUC can be inhibited by K^+ and β-bungarotoxin, which depolarize the cells and by hemicholinium-3.

Choline acetyltransferase (CAT) catalyses the conversion of choline and the other precursor, acetyl coenzyme A (acetyl CoA), to acetylcholine (Fig. 2.15). Acetyl CoA is derived from glucose in the brain. CAT is probably localized specifically in cholinergic nerve terminals and is a cytoplasmic enzyme, probably existing in multiple molecular forms. It does not appear that the enzyme is saturated with its substrate since precursor loading with choline can increase the rate of ACh formation in the brain. There are, however, no specific and potent enzyme inhibitors and it is unclear whether the enzyme plays an active role in regulating ACh synthesis and function.

The other precursor, acetyl CoA, is synthesized mainly in mitochondria and must therefore be 'transferred' in some way to the cytoplasm.

2.17 Metabolism

Difficulties in accurate assessment of the mechanism of the metabolism of ACh are, if anything, even more complex than those encountered in assessing synthesis. Undoubtedly ACh is hydrolysed by acetylcholinesterase but there are several esterases in the body and brain and these hydrolyse both ACh and other esters. The principal acetylcholinesterases are divided into two groups, 'specific' or acetylcholinesterase (AChE) enzyme and the 'pseudo' cholinesterase which is often referred to as butyrylcholinesterase (BChE) since it actively hydrolyses butyrylcholine and propionylcholine. AChE is probably mainly concerned with the breakdown of ACh which has been released from nerve endings. The role of BChE, which is much less active in this process is uncertain (see Silver, 1974) and a clear physiological role for it has yet to be demonstrated. It is found in glial cells, blood vessels, and plasma.

Cholinergic neurons all contain AChE and its distribution is therefore a reasonable guide to cholinergic areas. However, it has also been demonstrated in some non-cholinergic regions, such as hippocampus and cerebral blood vessels.

AChE is an extremely active enzyme and following release of ACh from the nerve ending there is a very rapid hydrolysis. The value of this is clear when it is realized that, in contrast to both choline and the amine neurotransmitters, no high affinity uptake system for ACh exists at the nerve ending. Once ACh is hydrolysed the choline formed can presumably provide a precursor pool for subsequent ACh synthesis (Fig. 2.15). One possible regulatory system is that the HAUC is inhibited by ACh thereby providing an inhibitory feedback loop.

The diagrammatic structure of AChE is shown in Fig. 2.16 and demonstrates the presence of the anionic site which attracts the positive ACh charge and the

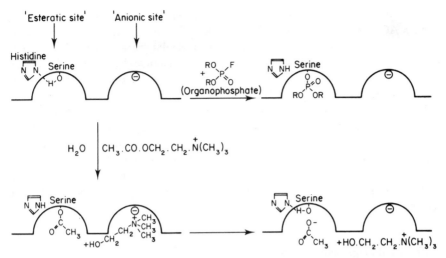

Fig. 2.16 The active site of acetylcholinesterase. The irreversible inhibition by an organophosphate at the esteratic site is shown together with the metabolism of acetylcholine

esteratic site about 5 Å away which binds the carbonyl moiety. The enzyme can be inhibited both reversibly and irreversibly. One reversible inhibitor is physostigmine (or eserine) which is an alkaloid and which has a much greater affinity for the enzyme that ACh. Nevertheless it gradually dissociates from the enzyme and it is thus reversible. Other reversible drugs include neostigmine and edrophonium, which has a very short duration of action.

The irreversible inhibitors include di-isopropyl flurophosphate and various other alkylphosphates. Their development is, in part, attributable to their potential as chemical warfare agents, and their value as insecticides. A common feature of the alkylphosphates is the phosphate grouping and it is this which produces the irreversible action of the drugs, since they phosphorylate the enzyme to a form in which it cannot hydrolyse ACh. On the basis of the structure of AChE proposed in Fig. 2.16, one antidote to the organophosphates has been developed, pyridine-2-aldoxime-methiodide (PAM), which is able to regenerate the phosphorylated enzyme (Wilson, 1958).

2.18 The acetylcholine receptor

The concept of more than one type of receptor for a single amine neuro-transmitter is reasonably new, but there is nothing novel in the general concept. The pioneering work of Sir Henry Dale in 1914 demonstrated the presence of two types of ACh receptor. The facts to support the concept of two, or more, types of ACh receptor can be stated simply as follows.

Atropine is a powerful antagonist of ACh actions on smooth muscles, cardiac muscle, and exocrine glands whilst having little effect on ganglia or skeletal muscle. In contrast d-tubocurarine inhibits the actions of ACh on skeletal muscle but not smooth muscle or ganglionic transmission. Since muscarine (an alkaloid) mimics ACh on smooth and cardiac muscle and exocrine glands but not ganglia whilst nicotine acts on ganglionic and neuromuscular transmission, the receptor types were therefore termed muscarinic and nicotinic by Dale. In addition hexamethonium blocks ganglionic but not neuromuscular transmission, suggesting a specific type of ganglionic receptor.

Much more is known about the structure of the receptor for ACh than that of any other major transmitter. The reason for this is perhaps that the electric organ of the electric eels (*Electrophorus*) is a very rich source of the receptor allowing high purification. The high specificity of certain snake venoms such as α-bungarotoxin for the receptor also assists in purification of the receptor, which is a nicotinic type. It is proving much more complex to isolate the muscarinic receptors.

2.19 Cholinergic agonists and antagonists

There are several cholinergic agonists, including various types of choline ester, which are structurally related to ACh but are more resistent to AChE. All are amines. Clinically the two most used drugs are bethanechol and carbachol, the latter having potent muscarinic and nicotinic actions. Pilocarpine in contrast is a tertiary amine and a muscarinic agonist only.

Several antagonists have been mentioned in Section 2.18. Muscarine antagonists include the alkaloids atropine and related drugs. Atropine and scopolamine have been used in medicine for centuries. Atropine is also called hyoscyamine, and occurs in the plant *Hyoscyamus niger*. The compounds are also known as belladonna alkaloids because they are found in *Atropa belladonna*, the deadly nightshade. In large doses the CNS effects of these compounds include excitation and mania-like effects and they have sometimes been used in Parkinson's disease (see Section 8.20).

Nicotine antagonists are the 'neuromuscular blockers' and include hexamethonium, nicotine, gallamine, *d*-tubocurarine, succinylcholine, and α-bungarotoxin. *d*-Tubocurarine is the active ingredient of curare, famous for its use on arrow tips by South American Indians. It acts by competing with acetylcholine for the receptor. Gallamine and α-bungarotoxin act in a similar way. Succinylcholine in contrast acts by depolarizing and then desensitizing the receptor.

2.20 Quantal release

The quantum release hypothesis derives from studies by Katz and colleagues on the mechanism of the release of ACh at motor nerve endings and states that the transmitter is released in discrete packets or quanta of transmitter. The technical problems in showing such a mechanism in the brain are formidable and it is not known whether it occurs. Furthermore the appropriateness of having such a release mechanism in the brain where much subtle modulation must occur, is questionable.

2.21 Criteria for a neurotransmitter

The criteria for identification of a neurotransmitter substance (Table 2.3) were originally stated by Chang and Gaddum in 1933, despite the apparent claims of many other subsequent workers. Again ACh was the transmitter under investigation and whilst the criteria can be shown to apply to ACh and its role in neuromuscular transmission there are formidable difficulties in proving a transmitter role for compounds in the CNS. Indeed it is doubtful whether some of the peptides now under investigation as possible transmitter substances (see this chapter, Part 5) would meet the criteria in any case.

Table 2.3   Criteria for the identification of a neurotransmitter

1. The compound should be present in neurons
2. There should be mechanisms present for its synthesis and release
3. There should be mechanisms present for its inactivation
4. Stimulation of neurons should lead to its release
5. Application of the authentic compound should mimic the effect of nerve stimulation
6. Inhibition of the inactivation mechanism should enhance the physiological effect produced by the transmitter suspect
7. Pharmacological antagonists should be equally active against the released transmitter and applied compound
8. Denervation should lead to receptor supersensitivity

List adapted from Chang and Gaddum (1933) and expanded.

PART 4   OTHER SMALL MOLECULES

2.22   γ-Aminobutyric acid (GABA)

The amino acid γ-aminobutyric acid (GABA) is now generally accepted to be a neurotransmitter. Nevertheless the functional role of this compound in mammalian CNS is only now beginning to be clarified.

It is a compound with a specific distribution pattern in the central nervous system and there is a good correlation between this distribution pattern and that of the enzyme that converts glutamate to GABA.

Studies on GABA are complicated by its intimate inter-relationship with carbohydrate metabolism. Glucose is a precursor of GABA and its degradation product is succinic semialdehyde which is metabolized to succinic acid, one of the compounds involved in the citric acid or Krebs cycle (Fig. 2.17).

GABA is therefore almost certainly involved in both neurotransmitter and

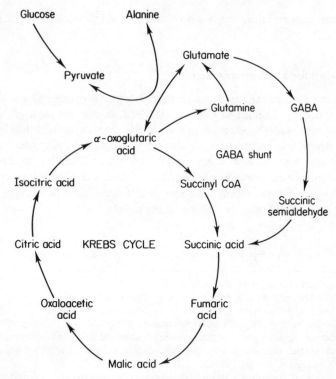

Fig. 2.17 Relationship between GABA synthesis and the tricarboxylic acid ('Krebs') cycle

metabolic systems. Furthermore, studies on this compound are complicated by the fact that its immediate precursor L-glutamate is probably also a neurotransmitter.

GABA is formed by the action of glutamic acid decarboxylase (GAD) on L-glutamate. This enzyme appears to be present specifically at the GABA nerve endings (Barber and Saito, 1976) and is pyridoxal phosphate dependent. GABA is metabolized by the action of GABA transaminase (GABA-T) which has a wide distribution and is also pyridoxal phosphate dependent. GAD can be inhibited by allylglycine, thereby leading to a decreased brain GABA concentration. GABA-T is inhibited by aminooxyacetic acid. Neither inhibitor is particularly specific which has complicated studies, as has the lack of specific GABA agonists and antagonists. Muscimol is probably the only agonist at present available. Bicuculline appears to be a GABA antagonist, acting at the receptor site and competing with GABA (Snodgrass, 1978). Picrotoxin and pentylenetetrazol (leptazol) also interfere with GABA neurotransmission but probably do so by acting 'beyond' the receptor. However, the interaction of all three drugs with GABA systems is both complex and still controversial. Benzodiazepines also interact in a novel way with GABA-ergic transmission, and this is discussed in Sections 4.5 and 4.6.

Another complication is that GABA is accumulated actively not only by neurons but also by glia, which makes distribution and metabolism studies more difficult, both to undertake and interpret.

One new technique for studying GABA has been that of measuring turnover by mass spectrometry. Developed in Costa's laboratory in Washington DC it involves the measurement of the rate of conversion of [^{13}C]-glucose to [^{13}C]-glutamate and [^{13}C]-GABA and the measurement of the latter two compounds by mass fragmentography (see Bertilsson and Costa, 1976). It is a technique which has produced valuable new data on the interaction of neuroleptics on brain GABA systems (for example Marco et al., 1976).

There is adequate evidence for GABA being an inhibitory neurotransmitter in crustaceans (see Cooper, Bloom and Roth, 1978) and this evidence has often been used to reinforce the argument that it acts as a transmitter in the mammalian CNS. Discussion of the evidence for GABA being a neurotransmitter may be found in several books (Roberts, Chase and Tower, 1976; Cooper, Bloom and Roth, 1978) but are beyond the scope of this book. Here we make the assumption that this amino acid fulfils a CNS neurotransmitter role and mention, in passing, main areas of interest.

There is now good evidence for a GABA–dopamine interaction (Lloyd, 1978). Much of this evidence is pharmacological and there is clearly a long striatonigral projection, lesions of the striatum leading to a marked decrease in GABA, GABA uptake, and glutamate decarboxylase (GAD) in the A9, A10 cell body area (see Hökfelt et al., 1977). Furthermore in an elegant study, Mao et al. (1978) showed that electrical stimulation of a specific region of the striatum led to an increase in nigral GABA turnover.

Increasing GABA concentrations produces changes in striatal DA which suggest that DA release has been inhibited and it has therefore been suggested that GABA acts presynaptically to inhibit DA release. However behaviourally it appears that GABA can also modulate DA-mediated responses postsynaptically (e.g. Cott and Engel, 1977) and is therefore apparently present in interneurons regulating DA neurotransmission.

The possible role of dopamine in the pathology of schizophrenia (Chapter 5) and Parkinson's disease (Chapter 8) and the involvement of GABA in Parkinson's disease and Huntington's chorea (Chapter 7) have stimulated interest in the use of GABA agonists and antagonists. Results so far have been disappointing but are discussed in the relevant sections.

Another area in which there has been much interest in GABA has been seizure disorders. Seizures can be produced in animals by various directly and indirectly acting GABA antagonists, such as bicuculline and picrotoxin and conversely can be prevented by increasing GABA concentrations with GABA transaminase inhibitors. The relationship between inhibition of seizures and brain GABA concentrations is not straightforward however, and seizures can occur at high GABA concentrations (see for example, Kuriyama, Roberts and Rubinstein,

1966). Good reviews are to be found by Meldrum (1975) and in Glaser, Penry and Woodbury (1980).

2.23 L-Glutamic acid

As stated in the previous section glutamate is the precursor of GABA and is present in high concentrations in the brain. It is also, of course, a compound involved in the intermediary metabolic systems of the brain.

Evidence for the neurotransmitter role of this amino acid is weak. It is an excitatory compound at certain invertebrate neuromuscular junctions and also in the brain of mammals. The problem, however, is that when iontophoresed into almost any region it has an excitatory action on unit discharge. This, together with the problems of mapping any compound involved in intermediary metabolism and showing it to have unequal distribution has led to suggestions that this compound produces a non-specific neuronal activation. The difficulties of defining a clear functional transmitter role for this compound are great and will not be considered further here.

2.24 Glycine

Glycine, like glutamic acid, is an amino acid. It is found in high concentrations in the spinal cord and may act as an inhibitory transmitter. It appears to have a specific distribution pattern. Strychnine is an antagonist at spinal glycine receptors and produces convulsions probably by glycine antagonism. Like glutamate, glycine is a known intermediary in energy metabolism but its metabolic paths are still speculative (Cooper, Bloom and Roth, 1978). Again a functional role is difficult to ascertain.

2.25 Histamine

Histamine is the product of the action of histidine decarboxylase on the amino acid histidine. Since the brain enzyme is unsaturated with respect to its substrate, histidine loading increases histamine concentrations in the central nervous system. The problem in investigating central histamine is that the diamine occurs not only in neurons but also in mast cells.

The major metabolic steps of histamine have been demonstrated (Fig. 2.18) and there have been some recent reviews outlining the increasing evidence for it having a neurotransmitter role in the CNS (Snyder and Taylor, 1972; Schwartz, 1975). Undoubtedly various psychoactive drugs alter the concentration of this compound in the brain (Snyder and Taylor, 1972) though the mechanisms of storage and release are poorly defined. The interest of this book in histamine is mainly because it has been suggested that it is involved in the action of antidepressant drugs. This idea was generated following the demonstration of a histamine-sensitive adenylate cyclase (Kanof and Greengard, 1978) and is discussed in Section 3.25.

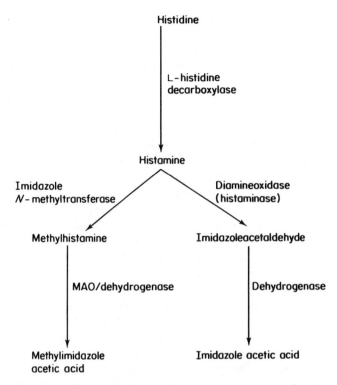

Fig. 2.18 The metabolism of histamine. Imidazole acetic acid can be conjugated to form the riboside which is excreted in significant amounts. The methyl imidazole acetic acid is also a major metabolite

PART 5   PEPTIDE TRANSMITTERS

2.26   General introduction

Until the mid 1970s psychopharmacological studies focused almost exclusively on monoaminergic systems with the addition of some studies on cholinergic, GABAergic and amino acid systems. Suddenly, however, an explosive growth of data has occurred; peptide neurotransmitters.

At present approximately 24 peptides have been proposed as possible central transmitters. In most cases the evidence is weak and absolute identification in the brain is still lacking in a few cases, due mainly to the problems of possible cross-reactivity between similar peptide compounds when using radioimmunoassay for identification.

Impetus to this area was undoubtedly given by the important studies on the identification and function of the enkephalins, and by the rather different endocrinological and gastrointestinal investigations being pursued at the same time. These investigations often contributed by supplying some of the immuno-assay methodology.

Two major questions arise. Why should many of the peptides be present in both the gastrointestinal system and brain, and how seriously can we consider peptides as putative neurotransmitters?

Embryology provides a possible answer to the first question. Pearse (1969) suggested that in the gastrointestinal tract there was the APUD (amine precursor uptake and decarboxylation) system. These APUD cells store a peptide hormone and a biogenic amine. As the gastrointestinal system and the central nervous system have the same embryological origins, the ectoblast (Pearse, 1969), the existence of the same peptides in both the gut and the brain is theoretically prob-able, and subsequent experimentation has confirmed this.

The second question is, at present, almost impossible to answer. The criteria for a neurotransmitter were enunciated by Chang and Gaddum in 1933 and have been listed in Table 2.3. It is not unreasonable to use them to see if any of the peptides can be confirmed as neurotransmitters. It immediately becomes apparent that the peptides cannot be neurotransmitters since we know almost nothing of the specific synthesis, storage, and release of the compounds, nor of their interactions with receptors. As will be seen perhaps enkephalin comes nearest to fulfilling the criteria, possibly because it has been the most extensively studied. In contrast some compounds (like TRH) have little effect in increasing unit activity when iontophoresed, although they may potentiate the actions of other transmitters.

There is now good evidence that several neurotransmitters coexist in the same nerve endings with various of the peptides. Examples are substance P with 5-HT or with TRH, and noradrenaline with enkephalin. A fuller list is given by Hökfelt *et al.* (1980). It seems that coexistence of amines and peptides may be a fairly general phenomenon* and this might be argued to contradict the generally accepted 'Dale hypothesis' which assumes that only one neurotransmitter is present at any one terminal. However, it is likely that the peptides are not con-ventional neurotransmitters but rather neuromodulators, regulating in some way the action of the neurotransmitters such as 5-HT and dopamine. Data suggest that no one peptide is associated exclusively with one classical transmitter. In different regions 5-HT coexists with either substance P or TRH. Equally, sub-stance P is present in neurons not containing 5-HT. It appears therefore that there are complex subgroupings perhaps allowing fine control of neuronal activity.

There is evidence for separate storage of the amines and peptides within the nerve ending. Reserpine at high dose has been shown to almost totally deplete monoamine stores but leave substance P concentrations unaffected (see the review of Hökfelt *et al.*, 1980).

We should not be surprised that the hypothalamic releasing hormones such as thyrotropin releasing hormone (TRH) are present in areas of the brain other than the hypothalamus. It is a peptide present in the brains of primitive animals without a pituitary gland. It is an historical accident that it was discovered first as a releas-ing hormone and this role is perhaps relatively 'recent'. Nature seems to use the same compounds in diverse ways, witness the role of 5-HT in the platelet as a

Table 2.4 Some possible peptide transmitters in the brain

Adrenocorticotrophic hormone	Melanotropin inhibiting hormone
Angiotensin	α-Melanocyte stimulating hormone
Cholecystokinin	β-Melanocyte stimulating hormone
β-Endorphin	Neurotensin
Met–Enkephalin	Somatostatin
Leu–Enkephalin	Substance P
Gastrin	Thyrotropin releasing hormone
Insulin	Vasoactive intestinal peptide
Leuteinizing hormone releasing hormone	Vasopressin
β-Lipotropin	

vasoconstrictive substance, in the gut as a factor involved in motility and in the brain as a transmitter probably concerned with sleep and mood.

A list of possibly active central peptides is presented in Table 2.4. However, in this section only a few will be considered namely the enkephalins and endorphins, TRH, MIF, substance P, ACTH fragments, and vasopressin.

2.27 Opiate peptides – introduction

The speed with which data are accumulating on the opiate peptides means that any review is out of date by the time of publication. However, it is possible to give a general review of this area, particularly with regard to its relevance to psychiatry. The review of the enkephalin data will be more detailed than that given to the other peptides, partly because much work has been done in this area, partly because generalizations made here probably apply to other possible peptide transmitters and finally because this work caught the imagination of most scientists.

2.28 Historical aspects of opiate peptides

There has been much work on the way that morphine and other opiates interact with neurotransmitter systems in the brain and many people felt that morphine was producing its central actions (catatonia, mood change, locomotor activity) by interacting in some way with neurotransmitters such as 5-HT and dopamine.

Opiates, however, were known to interact with both the guinea pig ileum and mouse vas deferens (see Kosterlitz and Waterfield, 1975), inhibiting the electrically induced contractions by inhibiting neurotransmitter release. This inhibition is blocked by the specific opiate antagonist naloxone and there is a good correlation between the clinical potency of the various opiates and their ability to inhibit these *in vitro* responses. It therefore occurred to Kosterlitz and Hughes (1975) in Aberdeen that there was perhaps a receptor present in these systems which was being stimulated by the opiates and which could therefore have an endogenous ligand.

There was other work consistent with this interpretation. Terenius (1973) in Sweden and Snyder and colleagues in Baltimore (e.g. Pert and Snyder, 1973) had shown that opiates would bind stereospecifically and with high affinity to central nervous system tissues. The binding sites had an unequal distribution and the binding affinity correlated with clinical potency. Furthermore electrical stimulation of the peri-aqueductal grey region of the brain produced analgesia, which was blocked by naloxone. It seemed reasonable, therefore, to suppose that the brain, on electrical stimulation could release an analgesic compound which would bind to receptors in the brain.

Whilst such statements may seem fairly obvious now, it is important not to underestimate the large conceptual leaps that were required to reach such conclusions.

2.29 The enkephalins

In 1975 the first report appeared on the structure of the endogenous opiates (Hughes *et al.*, 1975).

Two compounds were isolated, both pentapeptides, differing only in the terminal amino acid (Fig. 2.19).

These compounds were shown to act rapidly *in vitro* and their action is antagonized by naloxone. Their rapid metabolism means that it is difficult to show an analgesic effect except following modification of the molecule; a usual substitution is to put the non-physiological D-alanine in place of glycine (referred to as met-D-Ala^2-enkephalin); an antinociceptive effect can then be demonstrated.

Position in β-lipotropin : 61————————65

Met-enkephalin : Tyr-Gly-Gly-Phe-Met

Leu-enkephalin : Tyr-Gly-Gly-Phe-Leu

Fig. 2.19 Structures of the enkephalins

2.30 β-Endorphin

It did not escape the attention of Hughes *et al.* (1975) that the met-enkephalin structure is 'contained' in the amino-acid sequence of the pituitary prohormone β-lipotropin (β-LPH; Fig. 2.20). The fragment 61–91 known as β-endorphin (Fig. 2.20) has potent antinociceptive properties, marked affinity for the opiate receptor, and a long lasting action (Bradbury *et al.*, 1976), this last property probably being due to its resistance to enzymatic degradation.

There were suggestions made at the time of the discovery of β-endorphin that met-enkephalin was a 'breakdown product' of this longer peptide. However, studies showing unequal distribution of met-enkephalin and β-endorphin and studies which showed changes in met-enkephalin without changes in β-endorphin indicated that these two compounds have a separate existence.

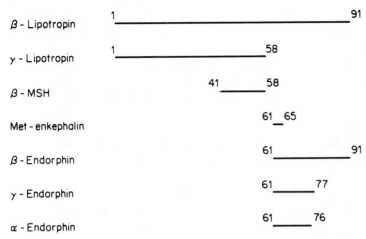

Fig. 2.20 The common peptide sequences of the enkephalins, endorphins, MSH, and lipotropins. See also Fig. 2.21 which shows the relationship of β-lipotropin to pro-opiocortin and ACTH

2.31 Multiple enkephalin receptors

In 1976 Martin and colleagues proposed the existence of three types of opiate receptor in an attempt to explain the spectrum of effects of the various opiate drugs. The receptors were designated μ (mu), δ (delta) and κ (kappa).

β-Endorphin has a high affinity for all three receptor types, whilst morphine has the greatest affinity for the μ-receptor and enkephalins for the δ-receptor although this is not true in all tissues examined (Lord *et al.*, 1977). Naloxone does not have an equal antagonist action at all three receptor types, being a rather better antagonist at the μ-receptor, which raises problems in investigating enkephalin function.

Martin *et al.* (1976) suggested that the hallucinogenic effects of certain opiates was the result of the interaction with the δ-receptor and this is the receptor that they suggested is involved with enkephalin–catecholamine interactions.

2.32 Distribution of enkephalins

Studies on distribution of enkephalins and endorphins are hampered by the lack of specificity of the radioimmunoassays. Nevertheless quite impressive studies have been performed which give us a clear idea of distribution.

The first point is that there are clearly separate β-endorphin and enkephalin-containing areas of the brain. β-Endorphin distribution studies suggest that there are cell bodies in the arcuate nucleus with projections to the midbrain and limbic structures.

Enkephalins, on the other hand, show a wide distribution with cell bodies and nerve processes occurring throughout the brain and spinal cord. High concentrations occur in midbrain, hypothalamic and dorsal horn of spinal cord, and there appears to be a relationship between enkephalin and aminergic areas of the brain. Certainly areas concerned with motor activity and behaviour are rich in enkephalin.

Of particular interest in terms of analgesia are the peri-aqueductal grey matter (PAG) and the descending 5-HT innervation of spinal cord. These areas are rich in enkephalins and there are data to suggest that there is a tryptaminergic regulation of the opiate receptors – decreasing 5-HT concentration antagonizes opiate analgesia whilst 5-HT reuptake inhibitors enhance responses (see, for example, Deakin and Dostrovsky, 1978; Uzan et al., 1980). Clearly, however, distribution is much wider than would be expected if the peptides were concerned solely with analgesia.

The ratio of met^5-enkephalin and leu^5-enkephalin varies in different regions but they nevertheless seem to share a common distribution pattern.

2.33 Synthesis, degradation, and release of enkephalins

Whilst met^5-enkephalin is 'contained' within β-lipotropin (β-LPH), it is clear that β-LPH is not the precursor of this small peptide. There is some evidence, however, that β-LPH is the precursor of β-endorphin. There is also some evidence to suggest that β-LPH is formed from a large precursor, possibly one containing both β-lipotropin and ACTH, named pro-opiocortin (Fig. 2.21). Data are now beginning to appear to suggest that met^5-enkephalin and leu^5-enkephalin are 'contained' within a large peptide precursor.

The inactivation of both β-endorphin and the enkephalins is almost certainly by enzymatic hydrolysis. Removal of the terminal group of enkephalin decreases activity by around 99% and aminopeptidases have wide distribution.

Enkephalins, like some other possible peptide neurotransmitters, demonstrate some of the conventional mechanisms involved in neurotransmitter release. For example, release can be evoked by potassium and is calcium dependent.

This area has recently been reviewed in several books (e.g. Beaumont and Hughes, 1979 and Kosterlitz, 1980).

2.34 Thyrotropin releasing hormone (TRH)

TRH was the first hypothalamic releasing hormone to be isolated and characterized (Schally et al., 1969; Burgus et al., 1970). It is a tripeptide having the structure pyroglutamylhistidylprolylamide. Development of reasonably specific radioimmunoassays allowed analysis of the peptide in the hypothalamus but also led to the observation that most TRH in the brain was extrahypothalamic (Jackson and Reichlin, 1979). TRH is present in animals devoid of a pituitary gland or thyroid stimulating hormone (TSH), indicating perhaps a general transmitter role, with specialization into a hypothalamic releasing factor occurring 'later' in evolution. A full review of the distribution and biosynthesis of TRH has been published recently (Jackson and Reichlin, 1979). All studies on TRH distribution using radioimmunoassay of course rely on the specificity of the TRH

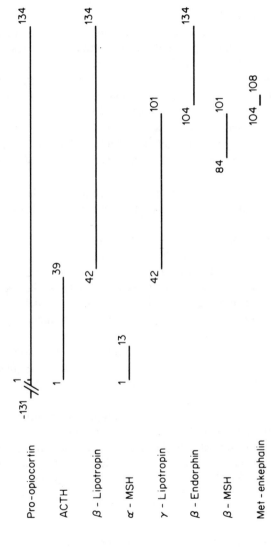

Fig. 2.21 Pro-opiocortin, ACTH, and endorphin structural relationships. See also Fig. 2.20 noting that the numbering of β-lipotropin differs when the ACTH–β-lipotropin fragment is considered, compared with when β-lipotropin alone is being numbered

antibody and this has been questioned (Youngblood *et al.*, 1978). Nevertheless, it is likely that the assayed compound, if not TRH, is nevertheless closely related. In this section the term TRH is used rather than the perhaps more accurate 'TRH-like'.

A few years ago it was suggested that the antidepressant action of the tricyclic drugs was potentiated by administration of tri-iodothyronine (T_3) the precursor of thyroxine (Wilson *et al.*, 1970).

One of the models for testing antidepressant potential in experimental animals is the 'dopa potentiation test' of Everett. In this model, tricyclics enhance the behavioural response which is seen after administration of a monoamine oxidase inhibitor (MAOI) plus L-dopa. In view of the T_3 data it was therefore thought worth trying the effect of TRH on the dopa potentiation test. It was discovered that low doses of TRH produced a marked increase of the MAOI/L-dopa behaviour. This increased response was seen in animals that had been thyroidectomized indicating that this action of TRH did not involve the thyroid–pituitary axis. TRH also enhances the behavioural effects of 5-HT and its agonists.

When large doses of TRH are administered peripherally the behavioural effect produced is rather similar to that seen after amphetamine. Recent work has explained this.

Direct administration of TRH into the mesolimbic forebrain (n. accumbens) causes the release of dopamine with all the attendant behavioural changes (Heal and Green, 1979). Biochemical data have demonstrated dopamine release *in vitro* (Kerwin and Pycock, 1979). Both studies suggested that the peptide did not cause dopamine release in the n. caudatus. Distribution studies on TRH have shown a high concentration in the accumbens but not caudate, suggesting that TRH may be having a physiological action.

It is unlikely that this action of TRH on dopamine release explains the effect on the dopa potentiation test, since the effects on dopamine release occur immediately whereas the effects on the potentiation test are apparent only after treatment at least 1–2 h earlier.

TRH has other interesting actions. For example, pretreatment with TRH shortens the sedation and hypothermia induced in rats by pentobarbitone or alcohol and this action is unrelated to the effects on the pituitary–thyroid axis. It has also been reported that TRH has anorexic activity and EEG alerting activity. In many of these actions TRH therefore resembles amphetamine.

2.35   Substance P

Substance P has a long history compared to most of the other CNS peptides which are currently creating interest. It was first reported and named in 1931 by von Euler and Gaddum who found a substance in both gut and brain capable of stimulating smooth muscle, and during the 1950s it was shown to have an unequal distribution in the CNS including the spinal cord. A major problem in working with the compound lay in the fact that the substance had not been fully purified or characterized, but in 1971 substance P was both identified as an undecapeptide (Chang et al., 1971) and synthesized by the same group (Tregear et al., 1971). This in turn allowed the raising of antibodies to the peptide and led to both an immunoassay and immunohistochemical mapping. Whilst we shall talk here about substance P, immunoreactive substance P would be more precise as closely related substances might also give a positive response in these techniques. Reports of the localization of substance P have recently been published (Cuello and Kanazawa, 1978; Ljundhl, Hökfelt and Nilsson, 1978; Hökfelt et al., 1980).

In the spinal cord substance P is present in primary sensory neurons as well as propiospinal neurons or spinal interneurons. It has also been reported in the vagus and taste neurons of cat and it has been proposed that substance P is involved in various sensory processes including pain (Henry, 1977).

Substance P is one of the peptides shown to coexist with the 'classical' neurotransmitters. Coexistence of substance P and 5-HT has been demonstrated as has the coexistence of substance P, TRH, and 5-HT. Nevertheless substance P also exists in areas where 5-HT is not present.

Iontophoresis of substance P in the spinal cord will lead to alterations in the rate of firing, but as yet little has been done to examine the function of this peptide in higher centres. One recent study, however, investigated the function of the pathway which originates in the medial habenula and innervates the ventral tegmental area. Of particular interest is the fact that the ventral tegmental area contains the dopaminergic A10 cell bodies which project to the mesolimbic, mesocortical forebrain.

When substance P is infused into the ventral tegmental area it produces locomotor activity which can be blocked either by injection of neuroleptics into the nucleus accumbens or by lesioning of the A10 pathway. It is possible therefore that substance P in the ventral tegmental area modulates the activity of the A10 system since it leads to release of dopamine in the nucleus accumbens (Kelley et al., 1979). Other studies have demonstrated that substance P injected into the substantia nigra can produce behavioural activation and circling when injected unilaterally. Furthermore neuroleptics decrease nigral substance P content. All these data point to dopamine-substance P interactions in the substantia nigra.

2.36 MIF or PLG

This peptide was named melanotropin release inhibitory factor because it appeared to inhibit the release of melanotropin (or melanocyte stimulating hormone). Subsequent work has suggested that this effect might have been due to impurities and MIF is now often referred to as PLG which stands for its tripeptide structure, Pro-Leu-GlyNH$_2$.

Following injection of radiolabelled PLG into the cerebral ventricles it can be shown that the label is widely distributed, but whether this indicates the specific distribution of the endogenous compound is uncertain.

In the dopa potentiation test (see TRH, Section 2.34) PLG, like TRH, enhances the behavioural response which follows administration of a MAO inhibitor and L-dopa. It has also been reported to antagonize oxotremorine induced tremor, a test for anti-Parkinsonian drugs, and Barbeau (1979) has reported that it is of value in treating Parkinson's disease (see Section 8.22).

2.37 The corticotrophin-related hormones

The compounds which have received most attention are the adrenocorticotrophic hormone (ACTH), peptide fragments, α-MSH and β-MSH (melanocyte stimulating hormone). Structurally there are distinct relationships between α-MSH and ACTH fragments (Fig. 2.21).

Most of the work on these compounds has been behavioural, the work stemming from observations that hypophysectomy leads to learning deficits in rats. It was subsequently found that learning could be largely restored by $ACTH_{1-10}$, $ACTH_{4-10}$, and α-MSH. None of these fragments had any adrenocorticotrophic activity (see for example De Weid, Witter and Greven, 1975).

2.38 Vasopressin

This peptide is an octapeptide released by the posterior pituitary.

It has been reported that vasopressin facilitates memory processes in rats (De Weid *et al.*, 1976) and this facilitation appears to be most effective when the peptide is given soon after the memory trial. It has also been reported that the compound improves human memory. A review on the effects of vasopressin on memory has recently been published (Kovacs, Bohus and Versteog, 1979).

Chapter 3

Depression and Mania

PART 1   PSYCHIATRIC ASPECTS OF DEPRESSION

3.1   General introduction

Depression is a feeling which most people have experienced at some time, and comprises those feelings of mood which lie on the sad side of a happy–sad continuum.

When mild it is a transient experience with no sequelae and is a normal often appropriate emotion. However, depression can be so severe as to be regarded as an illness causing severe distress, disruptive of life, and potentially fatal. This is the type of depression that is the subject of pharmacological research; however, at what point depression becomes serious enough to be regarded as an 'illness' is to some extent subjective.

Depression, being a disorder of mood or *affect* is categorized under the rubric *affective disorders*. This also includes *mania* and *anxiety states* (see Sections 3.21 and 4.1 respectively) which although recognizable as distinct entities have a close relationship to depression. This relationship will be discussed in the following sections.

3.2   Epidemiology

The prevalence of depression has been studied extensively both in terms of prevalence in the general population and the number of people seen by general practitioners or psychiatrists.

In Great Britain as many as 20% of the population have been found to have suffered from an episode of depression in the year preceding the enquiry. Although clearly categorized as depressed by reliable scales many of these 'cases' are only slightly above the threshold level and the most common social group is women aged 20–40 years. The presence of depression has been shown by Brown and Harris (1978) to be associated with life events, for example, bereavement, divorce, and moving house, and also with certain social factors, at least in women, viz.:

1. having several young children;
2. absence of a close confiding relationship;
3. lack of employment;

4. loss of mother by death or separation before the age of 11 years;

all of which are understandably distressing.

Not all of these patients are known to their general practitioners as depressed, and of those who are, only 10–20% are referred to psychiatrists. Studies have shown an annual referral rate for depression of around 300–400 per 100,000 population, of whom 10–15% would be admitted to hospital for treatment.

About half of these referrals are new cases and half people known to have suffered depression in the past. The patients referred to psychiatrists are of course, in general, more severely depressed than those found in the general population, and also differ in their demographic characteristics. They show a preponderance of people in middle life, women again being more common, with highest rates for the divorced and widowed.

3.3 Clinical features

Although the distinctions are to some extent artificial, it is convenient to consider the clinical features under several headings.

Depressed mood. The cardinal feature of depression is that of depressed mood; this fluctuates in milder depressions, but in more severe cases becomes constant and unresponsive to those stimuli which normally lift mood. The patient may be perplexed by this or accept it and become uninterested in life and pessimistic about the future. There is often considerable diurnal variation in mood with improvement as the day proceeds being the commonest pattern.

Some patients may hide their depressed mood whilst still exhibiting other symptoms, the so-called 'masked' or 'smiling' depressions.

The interpretation of altered mood may be confused by coexistent anxiety or overlap with an episode of mania (see Section 3.31).

Negative self-concept. A lack of self-confidence and loss of assertiveness may lead to the patient under-rating himself, feelings of worthlessness, self-blame and guilt. These may be quite irrational or delusional, and often involve ideas of financial ruin and hypochondriacal ideas, for example of suffering from cancer or venereal disease. Feelings of unreality and depersonalization may be present. Ideas of self-injury or need for punishment can be seen to stem from both lowered mood and guilt. Suicidal ideas are common, and may be freely admitted or concealed.

The above symptoms, being largely subjective, are difficult to quantify. Other symptoms may be assessed more objectively, either by direct observation or by measuring performance on simple tests.

Changes in motor activity. This may be either *retardation*: reduction in body movements, poverty of facial expression, sparse and slow speech, a general slowing of thinking, difficulty in formulating ideas, inability to make a decision; or *agitation*: restlessness involving pacing, fidgeting, the same thought recurring constantly without any action being taken on it.

Physiological symptoms. These are common and most typically consist of the following: disturbance of sleep, usually experienced as a tendency to wake early in the morning but sometimes difficulty in getting off to sleep; loss of appetite with resultant loss of weight; constipation and a feeling of being bloated may reinforce not eating; loss of libido or sexual drive with resulting impotence or frigidity; tiredness, muscle pain, difficulty in coping with physical tasks (which may lead to a patient giving up work).

Anxiety. The subjective experience of anxiety may be accompanied by the features of autonomic nervous system overactivity such as headache, sweating, trembling, and palpitations. Physiological symptoms may overlap with these symptoms. Anxiety as such will be considered in Chapter 4.

Mania. In some people episodes of depression alternate with episodes of mania. The clinical features are in some way the obverse of those seen in depression: distractability, speeded up thought and talk, many new ideas and a sense of well being (see this chapter, Part 4). The symptoms may in part overlap and for a time at least, confuse the diagnosis.

3.4 Classification

Some differences amongst depressed patients have been hinted at in the description of clinical symptoms, e.g. retarded versus agitated depression and the presence or absence of episodes of mania. Much attention has been paid to the existence of different subgroups. Such differences are not of only theoretical interest. 'Depression' is perhaps no more specific a diagnosis than 'chest pain', encompassing many entities of different aetiology which may therefore be capable of responding differently to different treatments.

The associations between prognosis and response to different treatment, and clinical features and other patient characteristics may be of great utility, leading through characterization of subgroups and the responses of these to different treatments to discovery of pathological mechanisms and aetiology. Aetiology remains obscure in psychiatry, although such a process can be seen to have taken place with regard to chest pain which could have its origin in heart, lung, stomach, etc. The analogy may appear crude but is apposite. That the various aetiologies are more obscure in psychiatry, does not argue against their existence.

The disentangling of such syndromes in depression is thus of fundamental relevance to treatment and research, particularly in psychopharmacology, where treatments remain empirical, and such theories as do exist about pathological mechanisms have often been suggested by the study of drug actions.

Conceptual issues. Criticisms of psychiatry are sometimes based on a confusion between aetiology and the mechanism whereby symptoms are produced. The psychopharmacological approach to depression has been criticized as inappropriate following evidence of the association between depression and social factors, such as early bereavement, socioeconomic group, age and size of family, and presence of close confiding relationships, as found by Brown and coworkers. The criticism may be simplified to 'She is depressed because she is underprivileged, to treat her with drugs is wrong and avoids the issue'. To treat the patient with drugs certainly will not itself change the social conditions, but whether or not it is wrong to so treat should be judged on outcome, not aetiology. A causal relationship does not show that there is no biochemical mechanism interposed between aetiological factors and the clinical manifestations.

Dementia following vitamin B deficiency has been shown to be related to social conditions. If the analysis of data is confined to the clinical picture and social factors, a 'causal' relationship can be shown, as for depression. More intensive investigation showed that the dementia is associated with vitamin B deficiency secondary to a general dietary deficiency, itself consequent upon poor social conditions. The 'treatment' is either vitamin B or improved social conditions (or in practical terms preferably both). Both work through the same mechanism. A similar analysis may be relevant for depression. There are many urgent arguments for improved social conditions other than the association with psychiatric illness.

The authors of such research refer to a causal relationship between social conditions and depression but it is not clear whether they believe their research demonstrates causality or that causality is the most likely explanation for the association found. A detailed analysis is not possible here, but it is worth reiterating that causality can never be 'proven' but can be shown to be increasingly likely. In practice these controversies are non-contributory as the pragmatic approach adopted by most psychiatrists would be to provide a broad based treatment, including drugs where appropriate, individual and family counselling, and an attempt to modify or reduce the impact of adverse social conditions.

Methodological issues.   Some of the clinical features described above are recognized subjectively, either by the patient or an observer, and mostly fall into the category of 'soft' data, incapable of precise and consistent measurement, and open to the interpretation and therefore biases of the assessor. This point is particularly important as most studies have been carried out on hospital populations, selected in the first instance by their general practitioner, perhaps because of some criterion such as having attempted suicide, or having broken the law as a consequence of their condition. They are therefore probably not a sample representative of the overall depressive population. To subject such data to sophisticated and powerful multivariate statistical techniques can be misleading and a bias held by the medical profession as to what constitutes depression may thus be dignified by analysis, and given apparent truth.

Clinical features are clearly of primary importance for diagnosis and management but any lasting value they have for classification purposes depends on their relationship to aetiology or pathological mechanisms. This may also be true for their predictive value, but is not so for such symptoms as suicidal ideation where the symptom itself may determine the development of the illness.

In addition to clinical data other data are obtainable, such as sex, age, family, history of illness, duration of illness (itself a measure as well as possible predictor of outcome!), marital state, housing condition, and other social circumstances. These are much more objective and are capable of being quantified more accurately.

Many attempts at classification and prediction of outcome have used a combination of such data, and where this improves reliability of diagnosis, may be worthwhile. However, the same clinical picture could be produced by more than one combination of different aetiologies (or mechanisms) and conversely the same aetiology (or mechanism) may be influenced by others to produce more than one clinical picture. Grouping together all types of data may thus obscure delineation of symptoms, and if two groups are found it may not be possible to decide whether the dichotomy is at the aetiological level, symptom level, or both, nor to identify the nature of the relationship between these. There is a case for restricting initial data to clinical data, and considering other data separately, a form of 'multiaxial' classification. Although used in child psychiatry, this is not used extensively in adult conditions, but a scheme proposed by Winokur will be discussed in Section 3.8.

Finally, it should be repeated that a classification should consist of classes which are mutually exclusive and jointly exhaustive, that is that every case diagnosed as depression should be allocated to one and only one class. Once this has been shown for the original group it should be replicated in an independent sample before the classification is accepted.

Findings and conclusions. The controversy over the results of studies on classification is fierce, not only with respect to the details but also on the relationship between classificatory groups. Suggested classifications fall into the following systems.

1. Simple, discrete classes, consisting of from one to five classes.
2. Tiered classifications, where groups may be divided into further subgroups.
3. Dimensional systems, with depressions being categorized as points on one or two continua or axes.

None of these proposals has achieved universal approval, but there are some areas of agreement.

3.5 Unipolar versus bipolar

One major difference is between those patients whose depressive episodes alternate with periods of mania (bipolar) and those who are free from manic symptoms (unipolar). Evidence that such a classification has some aetiological validity is given by genetic studies which have generally shown that relatives of bipolar patients have twice the likelihood of developing depressive illness, compared with those of unipolar patients. Furthermore the illnesses tend to 'breed true', that is, those relatives of bipolar patients who develop depressive illness are more likely to develop bipolar illness than the relatives of unipolar patients, and conversely those of unipolar patients are more likely to develop unipolar illness. Such as distinction, however, does not necessarily imply a categorical difference, as the various incidences could be explained as due to severity of illness.

A further difficulty is that it is not possible to say that a patient who is depressed will not develop a manic episode at some future time, although this likelihood reduces markedly after the occurrence of three consecutive episodes of depression.

3.6   Primary versus secondary

It is sometimes the case that depression arises in persons with pre-existing illness. This is of course likely to happen by chance in some patients but for many associations the likelihood can be shown to be much higher than that expected by a random interaction. This is true for the psychiatric illnesses, schizophrenia and alcoholism, and physical illnesses, rheumatoid arthritis, carcinomas, and endocrine disorders. It has also been shown that drugs may precipitate depression, e.g. hypotensive agents and steroids.

Hence a dinstinction between primary and secondary depression may be made, indicating a possible causal association. Again there are problems; if there is a pre-existing illness which is not diagnosed, some secondary depression will be classified as primary. Conversely, in common disorders some associations will be serendipitous and primary depression classified as secondary. Although attractive as a hypothesis there is little evidence that there is a difference between the two groups other than the existence of the association. As a research approach it has advantages but it has as yet provided little data regarding mechanism or treatment.

The significance of these above distinctions may be unclear, but there is general agreement about their descriptive validity.

3.7   Type A versus type B

Perhaps the greatest controversy in the classification of depressive illness is that surrounding the distinction between *psychotic* or *endogenous* depression, and *neurotic* or *reactive* depression.

The former is described as a severe illness of acute onset, and without obvious precipitants. There is a consistent lowering of mood with diurnal variation, guilt, and physical symptoms. The latter as a milder illness often precipitated by external events, varying in intensity as associated with symptoms of anxiety.

The use of the above terms is itself misleading. Endogenous and reactive presupposes that some depressions arise from within the person, and others are reactive to life stress. Such a difference is not borne out by studies of antecedent events. The terms psychotic and neurotic are perhaps the most confusing in psychiatry. They have no generally agreed meaning and are commonly used to dignify a global assessment of severity. It is arguable that they should be discarded but to do so would be to abandon terms capable of making a useful distinction which is central to psychiatric symptomatology: that of insight. That they are misused is not of itself an argument against them.

The heading to this section will have indicated a blunt but unprejudiced distinction between that of type A ('endogenous' or 'psychotic') and type B ('reactive' or 'neurotic').

There is general agreement that some distinction between types A and B is possible, but the nature of this distinction is more controversial. Most studies have used statistical techniques such as a multiple regression analysis, or multivariate analyses such as discriminate function analysis, principal component analysis, factor analysis, or cluster analysis, in an attempt to discriminate between the populations. The dubious validity of doing this on data of the quality available has been discussed above.

The main relationships suggested are:

1. two discrete conditions: the distribution of various scores is found to be bimodal with little overlap.
2. a hierarchal relationship. Type A is characterized by the presence of certain features, and type B by the absence of these. Other factors are common to both.
3. types A and B represent opposite poles of a continuous distribution which may be represented along a single continuum.
4. they represent a continuous distribution but one which requires two axes for its expression.

For practical purposes in psychopharmacological research considerations of the nature of the relationship are ignored and type A patients selected out of a depressed population, perhaps most similar to the hierarchy described in 2. This is perhaps a good compromise at our present stage of knowledge, as although most studies have identified a category corresponding to type A, there is much difference between their findings on type B, some identifying several clusters and some none.

In summary then it does seem possible to make some differentiations within the category of depression, although as yet these have not been shown to have any aetiological significance. Problems arise when different classifications are used, so it is clearly vital that studies should make their selection criteria explicit.

3.8 Classification according to family history

Although not conceptualized as part of a multiaxialclassification scheme Winokur (Schlesser, Winokur and Sherman, 1979) proposed a subtyping of unipolar primary depressive illness, according to family history of psychiatric illness. He proposes three types:

1. Familial Pure Depressive Disease (F.P.D.D.) in which a first degree relative is known to have suffered from depression, but no first degree relative has suffered from mania, alcoholism, or been labelled as an antisocial personality disorder.
2. Sporadic Depressive Disease (S.D.D.) in which no first degree relative has suffered from any psychiatric disorder.
3. Depression Spectrum Disease (D.S.D.) in which a first degree relative has suffered from alcoholism or been labelled as an antisocial personality disorder whether or not any first degree relative has suffered from depression.

That such a classification has some validity has been demonstrated by the dexamethazone suppression test, which may conveniently be considered in this section.

It is known that some depressed patients differ from the normal population in that their secretion of ACTH and, secondarily to this, cortisol, is not abolished by normal negative feedback mechanisms and this can be demonstrated by using dexamethazone as a challenge. Approximately 50% of hospitalized depressed patients will not suppress their cortisols in response to this whereas virtually all non-depressed people will.

The incidence amongst the subtypes above were found by Winokur to be:

F.P.D.D.: 82% 'suppressors';
S.D.D.   : 37% 'suppressors';
D.S.D.   : 4% 'suppressors'.

3.9 Quantification of ratings of depression

Although questionnaires have been developed to identify illness and provide a measure of severity at the same time, in psychopharmacology severity is usually assessed separately, and clearly must be when progress is being evaluated.

Attempts have been made to standardize diagnoses, either by structured interview or by operational criteria, as discussed above, and these have achieved high validity and reliability. However, it is often more difficult to assess degree than to make a judgement of presence or absence, especially when working with soft data, and attempts have been made to standardize measures of severity.

Depression rating scales have been developed of the following types:

1. observer rating scales;
2. self-rating by the patient.

Observer ratings. The most widely used scale is that produced by Hamilton (1960) – the Hamilton Rating Scale (HRS). The ratings are based on a clinical interview which aims to cover the preceding week. Ratings thus cannot be made more frequently than once a week. Seventeen items, including both mental and physical symptoms, are each scored on an ordinal scale, and the values summed to produce an overall rating. Inter-rater reliability is high (0.8–0.9) as is validity when measured by comparing scores with global assessment by a clinician. Other scales are variants of this approach.

The ratings may be biased by the beliefs or prejudices of the rater, but this is minimized by training and comparison of two raters' scores of the same interview.

Self-rating scale. Some of these are similar in style to the HRS, except that they are presented as a series of verbal or written questions which the patient himself rates. There are differences in the items used, and this is sometimes used to justify their use alongside observer scales. However, where they have been refined to correlate closely with the HRS, their use for providing information supplementary to this is dubious. Their general lower validity and reliability may increase the chances of false associations and those may be accepted, even when not confirmed by the HRS. They are also open to the biases or aspirations of the patient. Their advantage is that they are less demanding of professional time, and may be administered frequently.

Surprisingly good validity and reliability has been obtained using self-rated visual analogue scales where the subject is asked to make a mark on a line (usually 10 cm long) the extremes of which are marked, for example, 'happy' and 'sad'. Amongst other biases in these there is a tendency for patients to underestimate their deviation from the midpoint, a 'central tendency', which may preclude using a simple length as a measurement. The ease with which patients accept and understand these, however, may more than compensate for their apparent crudeness.

PART 2   BIOCHEMICAL ASPECTS OF DEPRESSION

3.10   General introduction

The current hypotheses on the biochemical basis of depression are the outcome of a fascinating development of ideas based mainly on observations of 'side effects' of drugs, logic, and serendipity.

For the past 25 years interest has been focused on the biogenic amine neurotransmitters. The development of hypotheses centring on the monoamines followed soon after the discovery of these compounds in the brain. Whilst it would be tempting to suggest that the continuing interest in monoamines indicates the strength of the original suggestions it seems more plausible that this interest is due to the fact that they were the first of the central neurotransmitters reported, together with the relative ease of studying and measuring such compounds.

Two general approaches can be used to investigate the biochemical changes which might occur in depressive illness. The first is to examine, for biochemical changes, tissue such as blood, cerebrospinal fluid, urine, platelets, and post-mortem brain tissue from patients suffering from depression. The second approach is to study the ways that drugs effective in treating the illness alter the biochemical function of the brain in experimental animals or biochemical parameters in tissues of depressed subjects. From these data extrapolation is made as to what changes may be occurring during depression. Both of these approaches have problems and these have been discussed in Chapter 1.

It has been mainly the 5-hydroxytryptamine system and the catecholamine neurotransmitter systems which have been implicated in depressive illness. In the latter system noradrenaline has received much more attention than dopamine. Until the relatively recent growth of data on neurotransmitter interactions in the brain the tendency has been for these monoamine systems to be considered separately and discrete theories developed. Increasing evidence that these two neurotransmitter systems are intricately inter-related suggests that the hypotheses are reasonably compatible, and this is discussed later (Section 3.30).

3.11   Historical aspects

In 1953 independent groups in Edinburgh and the USA reported that 5-hydroxytryptamine was present in mammalian brain tissue and within a year Woolley and Shaw (1954) proposed that this compound might be involved in the regulation of mood. This hypothesis was based primarily on the observation that 5-HT has a strong structural similarity to lysergic acid diethylamide or LSD (Fig. 3.1). It was already known that LSD was a compound which produced marked alteration of mood states. The view that 5-HT was involved in mood disorder received further support from clinical observations on the effects on the alkaloid reserpine. This drug was widely used in the 1950s to treat hypertension, but previously in India the plant *Rauwolfia serpentina* from which it is extracted had been used to treat mental illness for several centuries. It was noticed that

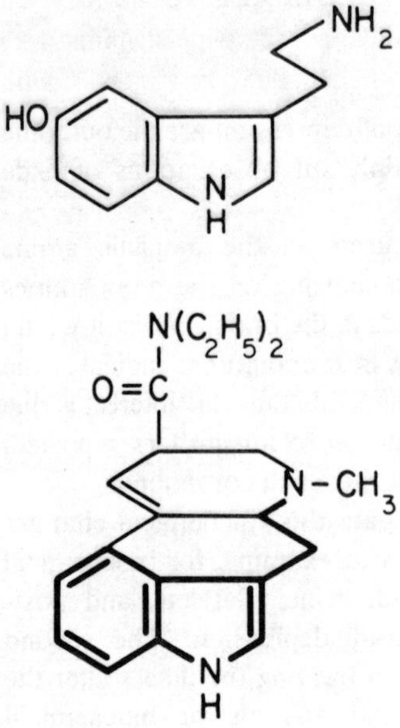

Fig. 3.1   The structure of 5-HT and lysergic acid diethylamide (LSD) with the similarity in structure outlined in heavy type

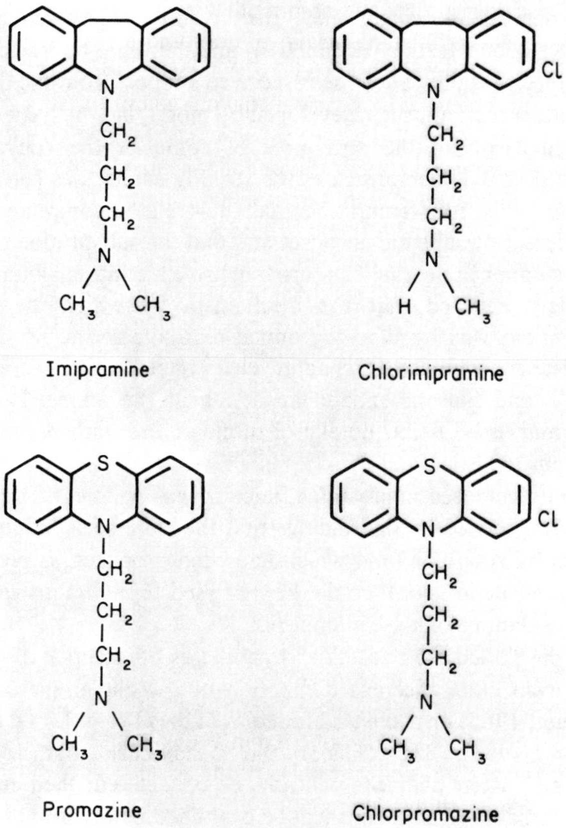

Fig. 3.2 The structural similarities between two tricyclics (imipramine and chlorimipramine) and two phenothiazines (promazine and chlorpromazine)

when used as an antihypertensive this drug could precipitate a severe depressive episode. Retrospective studies suggested that this occurred in up to 15% of patients. Subsequently reserpine was shown to deplete 5-HT (and also DA and NA) in rat brain.

During the early 1950s it was noted that some patients being treated for tuberculosis with the drugs iproniazid and isoniazid became euphoric. Subsequently Zeller (1959) demonstrated that these compounds inhibit the enzyme monoamine oxidase and therefore inhibit the degradation of the brain monoamines.

These two observations could obviously be linked in a hypothesis regarding the biochemical mechanism of mood disorders.

1. Depletion of monoamines (reserpine) produced depression
2. Elevation of brain monoamine concentrations (monoamine oxidase inhibitors) produced mood elevation.

Therefore depression might be the result of lowered monoamine neurotransmitter concentration (or function) and therefore be treated by elevating monoamine concentrations. Such ideas received support from the development of the tricyclic antidepressant drugs, developed by modifying the basic phenothiazine molecule. Comparison of the structures of some of the tricyclics with the phenothiazines (Fig. 3.2) demonstrates the striking similarities between these two classes of compounds. It is remarkable that such related compounds can be used for entirely different therapeutic purposes and that the substitution of two nitrogen atoms for one sulphur in the ring can turn a neuroleptic into an antidepressant.

It was widely believed that the mechanism by which the tricyclic antidepressants worked was by blocking amine reuptake at the nerve ending. This would be expected to increase the synaptic cleft concentration of transmitter (both 5-HT and NA) and thus potentiate the action of the amine. It is known that tricyclics will markedly inhibit uptake of monoamines both *in vivo* and in brain slice preparations.

The view that decreased monoamine function was an integral part of the mood change was strengthened by the findings that the state of sedation which can be induced in rats by reserpine (and which had been proposed as perhaps being an animal model of depression) could be reversed by administration of either tricyclics or monoamine oxidase inhibitors.

There thus developed the two major hypotheses of affective disorders: depression resulted from either decreased function of catecholamines, predominantly NA (Schildkraut, 1965) or decreased function of 5-HT (e.g. Lapin and Oxenkrug, 1969; Curzon, 1969). The clinical and behavioural data concerning the indoleamines and catecholamines will now be considered; then current ideas on the actions of antidepressant drugs will be examined before an attempt to put the observations together into a cohesive pattern.

3.12 Cerebrospinal fluid indoleamines

As stated earlier, interest in the involvement of 5-HT in depression stemmed initially from the suggestions of Woolley and Shaw (1954). The first direct indication for a possible change in central 5-HT function came when in a novel and provocative study, Ashcroft and his colleagues (1966) in Edinburgh reported that 5-hydroxyindole compounds in the CSF of patients were lowered in depression (Table 3.1). This initial finding led to a variety of studies on indole concentrations in the CSF. Unfortunately these data have failed to provide a clear result, some studies finding reduced concentrations of the 5-HT metabolite 5-HIAA in depressives, while others have failed to detect a change (see Green and Costain, 1979). Coppen's data unequivocally demonstrated decreased CSF 5-HIAA, his depressed patients having values half of his control group. Interestingly, these values did not return to normal on recovery.

Åsberg and her colleagues (1976) have presented data suggesting that 5-HIAA concentrations in depressives follow a bimodal distribution (Fig. 3.3) one group having normal and the other having low concentrations. People in the latter group

Table 3.1 Concentration of 5-hydroxyindoles in lumbar CSF from patients with psychiatric disorders

Diagnosis	n	Concentration 5-hydroxyindoles (ng/ml $\pm$ s.d.)
Controls (neurological diseases and alcoholics)	21	19.1 $\pm$ 4.4
Depressed	32	10.3 $\pm$ 3.8
Hypomanic	4	18.7 $\pm$ 5.4
Schizophrenic		
Acute	7	10.9 $\pm$ 2.3
Chronic	7	16.4 $\pm$ 2.9

Adapted from results of Ashcroft *et al.* (1966).

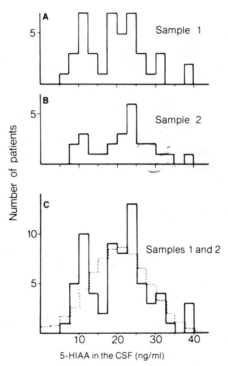

Fig. 3.3 Distribution of 5-HIAA in depressed patients: A, sample 1, $n = 43$; B, sample 2, $n = 25$; C, samples 1 and 2 combined. The dashed line represents the expected normal distribution (mean $\pm$ standard deviation $= 20.36 \pm 7.77$). The deviation from normality is significant ($\chi^2 = 19.76$ 9 d.f., $P = 0.02$). (Reproduced from Åsberg *et al.* (1976) *Science* **191**, 478–480. Copyright 1976 by the American Association for the Advancement of Science)

showed an increased tendency to commit suicide and her recent studies (Åsberg, Träskman and Thorén, 1976) have led to the suggestions that low CSF 5-HIAA may have a predictive value of suicide in that patients with a low CSF 5-HIAA show a greater predisposition to commit suicide. While it is true that Van Praag and Korf found a distribution skewed to the lower end in their study on CSF 5-HIAA content after probenecid (see below), many studies have failed to demonstrate the bimodal amine distribution, and clearly further data are necessary for the resolution of this question.

Several years ago Neff *et al.* (1976) demonstrated that probenecid administration to rats inhibits the acid transport system in the brain and thereby blocks 5-HIAA egress from the CSF. The role of 5-HIAA accumulation following probenecid thereby allows a rate of amine synthesis to be calculated (on the assumption that synthesis = metabolism = clearance of metabolite, in a single compartment system – see Section 2.12). This approach was developed by several groups, most notably by Van Praag and his department to study 5-HT turnover in depression. Various assumptions must be made and ethical considerations mean that only two determinations can really be made – one before and one after the probenecid. Full details of this approach are given in a book by Van Praag (1976). He discusses fully the several reports that 5-HIAA accumulation after probenecid is decreased in depressives. This would indicate that the rate of 5-HT synthesis is lower in these patients. In about 50% of the patients in longitudinal studies, the 5-HIAA accumulation returned to normal following clinical improvement. It was postulated that those patients who did not show a return to normal of 5-HIAA accumulation on clinical recovery might be those with an increased tendency to further episodes of depression.

Despite these data, there is now relatively little interest in CSF amine studies, for several reasons.

In the first place lumbar puncture is somewhat unpleasant, has small but definite risks attaching to it, and is therefore difficult to justify ethically for research purposes.

The second point is that there is still controversy as to the origin and significance of CSF 5-HIAA. While Garelis and colleagues (1974) have argued convincingly that 5-HIAA in CSF reflects central changes in 5-HT metabolism, several studies (e.g. Curzon, Gumpert and Sharpe, 1971) have demonstrated that in patients with a block in cerebrospinal fluid flow there is little decrease in the lumbar CSF 5-HIAA concentration. This would argue for a mainly spinal origin for the 5-HIAA, a conclusion also reached by Bulat and Zivkövic (1971) in their study on cats. Furthermore the lumbar CSF 5-HIAA concentration is related to the height of the patient, although of course this can be taken into account. A provocative finding by Curzon *et al.* (1980) was that depressives indeed have a lowered lumbar CSF 5-HIAA concentration (compared with controls). However, these same patients showed no decrease in ventricular CSF 5-HIAA content. This also argues against the lumbar CSF reflecting central amine changes, and it raises the possibility that lumbar 5-HIAA content reflects mainly the neuronal activity of spinal 5-HT systems. If this is so then lowered lumbar CSF 5-hydroxyindoles in depression may reflect mainly the motor retardation which is often present in these patients. This interpretation is given credence by the observation that increased motor activity raises CSF 5-HIAA content. It would not, of course, explain why in several studies showing lowered 5-HIAA in CSF the values did not return to normal on clinical recovery.

The next reason for the decline in interest in the CSF amine studies is that there can be a wide daily variation in values, and there is uncertainty as to whether the proportion of 5-HIAA passing directly into the blood stream is constant, thus making CSF values unreliable. There are also other methodological problems such as the specificity of the 5-HIAA methods (although this is not a problem in the recent studies using mass spectrometry) and the specificity of action of probenecid, since it can also alter plasma protein binding of tryptophan and thus influence the results obtained.

Comparison between studies has been hampered by the large differences in sampling techniques, times of sampling, methods of probenecid administration, and dietary intake.

The last reason for the decrease in the number of studies investigating CSF amines is the significance of the data in terms of assessing amine function. Even if the 5-HIAA in lumbar fluid reflects central amine changes, what does it mean in terms of furthering our understanding of the pathology? It is reasonable to assume that a proportion of the 5-HIAA being measured results from intraneuronal metabolism of 5-HT. Thus only a proportion of the measured metabolite has been released, produced a physiological response (stimulated the receptor), and been metabolized. Animal data have now clearly demonstrated that changes in monoamine transmitter concentration and synthesis may occur without necessarily altering 5-HT function. Thus whilst it may be reasonable to assume that amine metabolites reflect *metabolic* changes it must not be assumed that this is the same as a functional change. This view is perhaps supported by the data of Coppen (1973) which showed mood improvement occurring without a corresponding metabolic change.

Finally it is perhaps worth saying that despite all these difficulties and methodological objections, the overall data obtained do point to the possibility of a decreased 5-HT concentration or metabolism in depression.

3.13 Post-mortem brain studies

If there is decreased 5-HT concentration or synthesis in depression then one might hope to see changes in brain 5-hydroxyindoles in post-mortem tissue from depressives and there have been various studies examining this possibility. Two studies in the late 1960s reported that the concentration of 5-HT in the brain stem of suicides was lowered, and a further study indicated that the 5-HT content was not lowered, but 5-HIAA was decreased. Lloyd *et al.* (1974) have also reported decreased 5-HT in the raphe nucleus but not in other 5-HT areas. This group also obtained some indication of lowered 5-HIAA concentrations.

Again this work has been widely quoted in support of 5-hydroxyindole changes, but it is necessary to question the data. One assumption made in some of these studies was that a suicidal death is the culmination of a recorded or unrecorded case of depression, as suggested by Sainsbury (1968). However, this is a controversial idea and has been contested. Certainly, the view of Traskman *et al.* (1981) is that suicide and depression are not necessarily related and that it is predominantly the suicide cases who have lowered 5-hydroxyindoles in the CSF. If this view can be supported then it may go some way to explaining some of the conflicting post-mortem data.

There are also other problems inherent in post-mortem studies; the handling of the tissue between time of death and freezing of the brain, the time of death, both time of day and time of year and type of death (suicide and method, pneumonia etc.) all affect the measurements. It has taken a long time to establish unequivocal post-mortem changes in schizophrenics (see Section 5.10) and one would like to see similar care applied to a study of post-mortem biochemistry in a group of depressives especially in view of the more recent studies of Beskow *et al.* (1976) and Cochran, Robins and Grote (1976), both of which failed to detect any 5-hydroxyindole differences between suicides and controls.

3.14 Plasma tryptophan

In Section 2.3, the evidence for the dependence of brain 5-HT synthesis on brain tryptophan concentration was discussed and it was also shown that brain tryptophan content was determined, at least in part, by peripheral tryptophan availability. However, it has been only relatively recently that interest in possible changes in peripheral tryptophan availability has been examined in depression. The reason for this is that it has been generally felt that only large 'non-physiological' changes in plasma tryptophan would provoke changes in brain 5-HT synthesis. The demonstration by groups such as that of Fernstrom and Wurtman (see their review, 1974) that small changes in plasma tryptophan concentration could alter the rate of brain 5-HT synthesis suggested that perhaps small changes in plasma tryptophan might lead to alterations in 5-HT synthesis and thus precipitate or potentiate a mood change.

Clinical research was further stimulated by the work of Curzon and his colleagues (see Section 2.3) which demonstrated that a major determinant of tryptophan availability to the brain was that function of the tryptophan in the plasma which was 'free', that is not bound to plasma albumin. It therefore seemed reasonable to examine whether there were alterations in the plasma tryptophan content in depression.

In fact the results of experimentation in this area have not been encouraging and in keeping with much other data on biochemical changes in depression, the results obtained are inconsistent.

Coppen, Eccleston and Peet (1973) reported that depressed women have lowered concentrations of free but not total tryptophan in their plasma. Similar findings have been reported by some other groups but equally there are other well controlled trials which have failed to demonstrate any alteration in plasma tryptophan concentrations either free or bound, in depression (see review of Green and Costain, 1979). Furthermore if there were changes in peripheral tryptophan availability this should be detected in examination of central tryptophan content.

It is unclear what CSF tryptophan reflects in terms of tryptophan availability for the brain since much tryptophan must be present for protein synthesis and one can speculate that the important tryptophan is only that taken up and present in nerve endings. Nevertheless on the assumption that CSF tryptophan and intraneuronal tryptophan are related it is interesting to note that there is at least one report of lowered CSF tryptophan in depressives although here again there is other work failing to confirm this.

As shown in Section 2.3 most tryptophan is metabolized peripherally and only a small proportion enters the brain. The first enzyme on the quantitatively major pathway tryptophan pyrrolase (see Fig. 2.3) has its activity increased by administration of its substrate tryptophan or corticosteroids such as corticosterone and hydrocortisone. There is evidence that corticosteroid production in depression is often higher than normal. It has been proposed therefore that this increase in circulatory corticosteroids increases the activity of hepatic pyrrolase and thus increases tryptophan metabolism down this peripheral pathway. The result of this would be to decrease the availability of tryptophan for the brain (for example see Curzon, 1969). Certainly corticosteroid administration to rats does increase pyrrolase activity and decrease plasma tryptophan (both free and bound) and, in probable consequence decreases both brain tryptophan and 5-HT synthesis (see Green, 1978). It seems unlikely, however, that such changes in tryptophan metabolism occur in depression. This is suggested by the failure to find consistent alterations in plasma tryptophan content in depressives. Furthermore, while there have been reports of increased urinary excretion of kynurenine pathway metabolites (namely kynurenine and 3-hydroxykynurenine) from several laboratories, these data have not been confirmed in other departments, and there is now doubt about the specificity of the kynurenine assay used in many of these studies.

3.15 Oral contraceptives, tryptophan metabolism, and depression

A further contentious area of tryptophan metabolism involves the effects of oral contraceptives. It has been stated for some time that there can be a mood change in women on oral contraceptives (OC) and that this is the result of induction of tryptophan pyrrolase by the OC, thereby increasing metabolism of tryptophan down the kynurenine pathway (Fig. 2.3) and leading to a relative deficiency on the 5-HT pathway. This theory has rested on two main observations:

1. the urinary excretion of certain kynurenine pathway metabolites is increased in women on OC; and
2. some animal data indicating that OC administration may increase pyrrolase activity (these data are dubious in view of the failure to observe similar changes in pregnant animals).

Table 3.2 Total urinary excretion of various tryptophan metabolites by control subjects and those on oral contraceptives (OC). Results expressed as mean ± s.e. mean of the excretion by the nine control subjects and fourteen OC subjects. Results show total μmol metabolite excreted during collection period stated in table

Metabolite measured	Group	Collection times			
		Control period 10.00 h–10.00 h (24 h)	Experimental period		
			10.00 h–15.00 h (5 h)	15.00 h–10.00 h (19 h)	Total (24 h)
Kynurenine	Control	22.75 ± 1.56	68.52 ± 18.14	39.62 ± 17.91	108.11 ± 22.42
	OC	4.08 ± 0.52*	39.36 ± 7.58	22.45 ± 8.63	61.82 ± 11.11†
3-OH kynurenine	Control	ND	212.09 ± 29.37	356.38 ± 41.52	568.48 ± 52.45
	OC	ND	539.15 ± 97.32§	448.66 ± 92.27	987.8 ± 168.12
Xanthurenic acid	Control	ND	9.27 ± 2.87	39.70 ± 6.63	48.97 ± 6.09
	OC	ND	31.31 ± 11.46	96.82 ± 25.41†	128.14 ± 25.07‡

Experimental period in the 24 h following an oral tryptophan load (50 mg/kg).
Different from respective control: *$P < 0.001$; †$P < 0.05$; ‡$P < 0.025$; §$P < 0.01$; ND, not determined.
Taken from Green *et al.* (1978) with permission of Macmillan Journals Ltd.

Table 3.3 The plasma half-life ($T_{\frac{1}{2}}$), plasma clearance, and apparent volume of distribution (V_d) of tryptophan following an oral tryptophan load (50 mg/kg)

Group	$T_{\frac{1}{2}}$ (h)	AUC (μg/ml per h)	Plasma clearance (ml/kg per min)	Apparent V_d (l/kg)
Control	2.12 ± 0.15	291 ± 30	0.716 ± 0.082	0.572 ± 0.074
OC	2.20 ± 0.13	330 ± 18	0.792 ± 0.057	0.488 ± 0.027

AUC, area under plasma tryptophan versus time curve. Results show mean ± s.e. mean of determination on the ten control subjects and fourteen subjects on oral contraceptives (OC). Taken from Green *et al.* (1978) with permission of Macmillan Journals Ltd.

In a recent study it was indeed confirmed that OC users excrete increased amounts of certain tryptophan metabolites (Table 3.2). On the other hand no evidence was obtained for OC changing the rate of tryptophan metabolism through the pyrrolase step (tryptophan plasma half-life, clearance, and volume of distribution being very similar in OC users and drug free controls – Table 3.3). It was concluded (Green *et al.*, 1978) that the changed metabolite excretion pattern might be due to a relative vitamin B_6 deficiency since some of the metabolizing enzymes are B_6 dependent. A vitamin B_6 deficiency has certainly been noted in OC users and one study showed that in a group of women on OC suffering from mood change, a proportion had alleviation from the depression following pyridoxine administration (Adams *et al.*, 1973). Whilst 5-HTP decarboxylase is B_6 dependent, it seems extremely unlikely that a deficiency would have a marked effect since this enzyme has high activity in the brain and in animals significant enzyme inhibition can be achieved with little effect on monoamine function.

In any case, many workers do not now believe that OC use precipitates mood change and cite psychological rather than pharmacological causes as the origin of the reports (see for example Weissman and Slaby, 1973) and one recent clinical pharmacology textbook totally discounts the view that oral contraceptive use produces depression (Bochner *et al.*, 1978).

3.16 Urinary 5-HIAA

In early experimental work on indoleamine metabolism in depression, studies were made of urinary excretion of 5-HIAA. Various changes in excretion patterns were reported. However, in view of the small fraction of 5-HIAA in the urine that originates from the CNS it does not now seem worth paying too much attention to this work. Most of the 5-HIAA presumably results from 5-HT metabolism in the gut.

Whilst it would be useful if peripheral 5-HT metabolism reflected central metabolism there are good reasons for thinking that this is not true. For example, tryptophan loading (that is, giving a patient a high single dose of tryptophan) can be shown to increase central 5-HT synthesis in animals and result in increased CSF 5-HIAA in humans but does not alter urinary 5-HIAA excretion at all.

3.17 The blood platelet

The desire to have some peripheral 'marker' of central 5-HT systems has resulted in much attention being focused on the blood platelet, for the following reasons: it is a simple system, obtained easily by blood sampling, and contains 5-HT. Not only do platelets contain 5-HT, but the amine is stored in granules accumulated by an active uptake system, and degraded by MAO. Several workers have therefore suggested that the platelet might be a useful model for the central serotonergic neuron (Pletscher, 1978).

The ease of studying MAO activity in platelets has led to a variety of studies examining the activity of this enzyme in the platelets of depressed patients. However, data obtained have not been encouraging and no clear evidence for a change has been seen (see Pletscher, 1978).

There does, however, seem to be some evidence for an alteration in the active uptake of 5-HT into the platelets of depressed subjects. Tuomisto and Tukianen (1976) observed that active uptake was decreased in these patients compared with controls, a finding subsequently confirmed by several other groups. Furthermore this change was less on recovery and was not brought about merely by the presence of psychoactive drugs in the plasma.

The recent observations on the presence of 5-HT receptors on the platelet membrane and their alteration by drugs (see Peters, Elliot and Grahame-Smith, 1979) suggests that the next era in the use of the platelet in psychopharmacology will be investigation of these receptors in psychiatric illness and during drug therapy. (The section on tricyclic binding (Section 3.26) is also relevant to this.)

3.18 Catecholamine metabolites and cerebrospinal fluid

The support (and thus the most research) for the indoleamine hypothesis seems to have been strongest in the United Kingdom and Europe whilst the catecholamine hypothesis has received the strongest support from the USA. This is probably historical, the hypotheses having been stated, at least in their most basic forms, respectively in Europe and the USA.

Approaches to the demonstration of possible changes in catecholamine function have been very similar to those used to examine indoleamine metabolism.

Arguments about the significance of cerebrospinal fluid measurements of catecholamines are very similar to those raised about CSF indoleamine metabolites (Section 3.12), and our objections to the validity of such data in elucidating central monoamine function must therefore remain. With regard to homovanillic acid (HVA), the acid metabolite of dopamine, it appears that this compound reflects brain dopamine metabolism. There is a gradient of HVA down the cord and partial or total blockade of the subarachnoid space results respectively in a decrease or disappearance of HVA from the CSF below the block. Garelis et al. (1974) have suggested that HVA originates mainly from the caudate nucleus. It had been generally felt that the spinal cord contains few dopaminergic nerve endings, but recently Neff, using mass fragmentography (quantitative gas chromatography – mass spectrometry) has presented elegant data showing unequivocally that there are dopaminergic projections from the striatum to the cord (Neff et al., 1980).

Several investigations have found that lumbar CSF HVA concentrations are decreased in depressed patients although the difference failed to reach statistical significance in many studies (reviewed in Green and Costain, 1979).

One report suggests that HVA increased in lumbar CSF on recovery, but this is in contrast to several others which have not found any HVA increase on mood improvement (as in the case of 5-HIAA) and these observations would argue against the lowered HVA being due purely to motor retardation.

There are also reports that probenecid induced accumulation of HVA is lower in depression, indicating a decreased dopamine turnover (see the review of Van Praag, 1976). However Van Praag and colleagues have suggested that this change is not due to the mood change but rather the motor retardation seen in depressives. Against this interpretation, however, is the failure of HVA to increase on clinical recovery reported by Sjöstrom and Roos (1972), and further the work of Roos and Sjöstrum (1969) amongst others who found lowered HVA (and incidentally 5-HIAA) in groups of manic patients. Van Praag has pointed out that with two distinct sites of dopamine function – the nigrostriatal and mesolimbic, controlling respectively (it is postulated) motor function and affect – the interpretation of such metabolite data is complicated.

The origins of the noradrenaline metabolite 3-methoxy-4-hydroxphenylglycol (MOPEG) are unclear. There appears to be no gradient of the metabolite down the cord. MOPEG concentrations in the CSF of depressed patients have been reported to be decreased in some studies whilst other investigations have found normal concentrations (see Green and Costain, 1979).

It is difficult to assess these studies since MOPEG occurs both free and as a sulphate ester in the CSF and in many studies only the total changes were examined. It has also been suggested that cordal NA metabolism can account for a distinct proportion of the lumbar MOPEG present. There is also another NA metabolite, the acid metabolite vanillylmandelic acid (VMA) and this has also been reported to be lowered in depression.

In contrast to these various 'positive' findings in CSF such limited data as are available on post-mortem brains suggests there are no significant changes in the concentration of NA or DA or enzyme activities in suicide post-mortem brains. Two of these studies were ones in which changes in 5-HT or 5-HIAA concentrations were observed.

3.19 Plasma, platelet, and urinary catecholamine studies

Plasma has been little studied although there is one report that plasma tyrosine is unaltered in depression. The reason for this lack of interest (cf. plasma trypto-phan concentrations – see Section 3.14) is that administration to animals of either phenylalanine (which is converted to tyrosine in the liver) or tyrosine fails to alter catecholamine synthesis.

Platelets on the other hand have received much interest. They will transport dopamine actively but this system has been shown unequivocally to be the same one as that which normally transports 5-HT. Although there is clearly no advantage to studying both 5-HT and dopamine uptake, several trials have done just that! Nor does the system have any value as an index of 'physiological' dopamine transport changes.

Attention has therefore been concentrated on changes in platelet MAO activity. However, again data have been inconsistent, there being reports of both decreased and increased MAO activity in depressives. Schildkraut *et al.* (1977) have attempted to correlate platelet MAO activity with MOPEG excretion in depressives. They suggested that a group of depressed patients with 'bizarre antisocial behaviours' had high platelet MAO activity and low MOPEG excretion, but the number of patients was very low. Furthermore our own data in experimental animals have shown that there can be marked peripheral MAO activity changes with little change in the central enzyme activity (Youdim *et al.*, 1980). It seems extremely unlikely that the activity of such a ubiquitous enzyme as MAO – located as it is on the outer mitochondrial membrane – is necessarily going to change similarly in both the platelet and brain.

The likelihood of platelet MAO reflecting brain MAO activity in depression is remote and there are now few studies on platelet MAO activity taking place, other than as an index of inhibition of the enzyme following MAO inhibitors, where the measurement is, of course, justified.

It appears that in several species, including man, MOPEG or the sulphate ester is the major NA metabolite in the brain. It has been suggested that in urine the major proportion of the MOPEG derives from this central metabolism of NA, the metabolite of peripheral NA metabolism being VMA. While there are several reports of lowered MOPEG excretion in depression, these studies demonstrated that not all patients excreted less MOPEG and there are also reports failing to confirm these findings.

Nevertheless, several groups (e.g. Schildkraut *et al.*, 1977) have found that bipolar manic depressives and schizo-affectives excrete less MOPEG than controls, whereas unipolar endogenous depressives did not show this decrease.

All these data therefore point towards abnormal catecholamine metabolism in some depressed patients. The value of such data is considered, after the pharmacological observations, in Section 3.30.

3.20   *In vivo* studies on neurotransmitter function

An indirect method of assessing the potential activity of neurotransmitters has been developed using specific neurotransmitter agonists to stimulate a rise in one of the pituitary hormones – the so-called neuroendocrine challenge tests. Many of the pituitary hormones are known to be under the control of the neurotransmitters considered in this chapter: for instance dopamine is known to stimulate growth hormone (GH) release, and inhibit prolactin (PL) release, and noradrenaline to stimulate GH release through α-adrenergic activity. In the past many studies used insulin, amphetamines, or L-dopa as challenges but the standardization of these is difficult, and as they have multiple actions the interpretation of any finding is further complicated.

More recently studies using clonidine (an α_2-adrenoceptor agonist) as a challenge have suggested that there is an alteration of the response of GH in patients with type A depression but not type B. Whether this is of aetiological importance, or is due to a secondary hypothalamic disturbance in type A depressives remains to be elucidated.

PART 3   THE PHARMACOLOGY OF DEPRESSION

3.21   Historical aspects

The earliest of the current treatments to be used was electroconvulsive therapy (ECT), introduced by Cerletti and Bini (1938). Medicine had long had an interest in electricity (Fink, 1979) but the reasons for the experiments of Cerletti and Bini were due, not to an interest in electricity *per se*, but rather based on the belief, now known to be fallacious, that schizophrenia and epilepsy were mutually exclusive. Furthermore it is now generally recognized that ECT is not a useful treatment for schizophrenia, but its usefulness in depression emerged with clinical experience. Whilst it is not a pharmacological treatment it will nevertheless be examined in this chapter, since it is reasonable to assume that it is working through neurotransmitter systems. It is interesting that this treatment for depression, which has been used so widely, is perhaps one of the least understood biological psychiatric treatments and the least critically assessed. This is doubtless due to the fact that scientific methodology in the form of controlled trials came to medicine after ECT had become established.

The second advance in the pharmacological treatment of depression was the advent of monoamine oxidase inhibitors, with the tricyclic antidepressants, developed from the phenothiazines (see Section 5.13) being released commercially soon after. Recently a new generation of putative antidepressants has appeared. some being cyclic compounds but others having no clear structural similarity to existing antidepressant drugs. Some have been developed as relatively specific reuptake inhibitors (e.g. fluoxetine, which inhibits 5-HT reuptake, with little effect on NA systems) while others pose the interesting problem that they are clinically effective antidepressants but have little effect on reuptake systems (e.g. iprindole and mianserin). This section considers all these treatments, together with L-tryptophan (with and without concomitant MAOI administration) and their possible modes of action.

3.22   Monoamine oxidase inhibitors

The story of the MAO inhibitors began with the observation in the USA that iproniazid, an antituberculosis drug, had a euphoric effect in some patients. The inhibitors were introduced into the UK in the early 1960s. However, conflicting data soon appeared concerning their efficacy. Particularly important was the

Table 3.4 Efficacy of various antidepressant treatments

		% Improved		
Treatment	n (M/F)	Male	Female	Total
Placebo	51 (15/36)	53	42	45
Phenelzine	50 (15/35)	60	29	38
Imipramine	58 (22/36)	82	67	72
ECT	58 (21/37)	71	92	84

n is total number of patients investigated with the male/female numbers shown.
Adapted from Medical Research Council (1965).

MRC trial of antidepressant treatments (Medical Research Council, 1965), the results of which are summarized in Table 3.4. It can be seen that phenelzine was less effective than placebo in treating severe depressive illness and it has been argued MAO inhibitors are effective in 'type B' depression or anxiety states rather than severe depression; but again the data on this are conflicting.

While information on the efficacy was being published, reports also started to appear on a severe drug induced reaction or side effect. This is the so called 'cheese reaction', a hypertensive crisis which can occur in patients who have ingested certain foods (Table 3.5). These foods have in common the fact that they contain a high concentration of tyramine, which, because it is no longer metabolized by MAO, is taken up by peripheral vascular adrenergic nerve endings, displaces NA, and thereby produces the hypertensive reaction (see the review of Marley, 1977).

As a result of these problems and the reports that some of the drugs were hepatotoxic, there was a rapid reduction in the use of MAOIs in clinical practice.

There are several MAO inhibitors available (Fig. 3.4). With the exception of tranylcypromine, they are hydrazine derivatives related to the 'parent' compound iproniazid. Tranylcypromine is known to have some 'amphetamine-like' actions

Table 3.5 Some of the common foods that can produce a hypertensive crisis when ingested by people taking monoamine oxidase inhibitors

Cheese (particularly Camembert)
Wines (particularly Chianti)
Bovril ⎫
Marmite ⎭ Yeast extracts
Chicken livers
Broad beans
Avocado pears
Pickled herrings
Beers
Yoghourt
Flavoured textured vegetable protein

Phenelzine

Iproniazid

Nialamide

Tranylcypromine

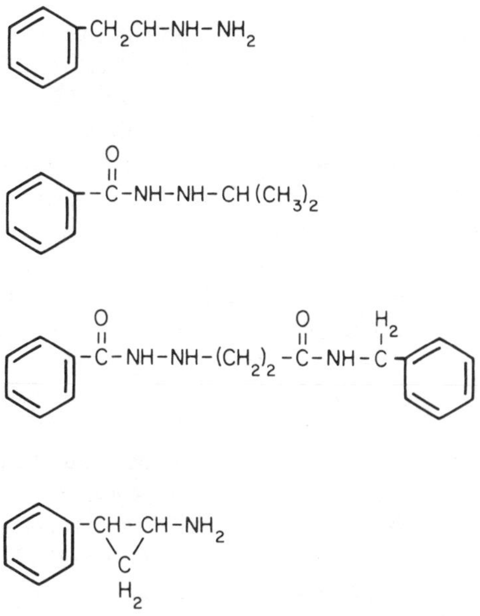

Fig. 3.4 The structures of various monoamine oxidase inhibitors

Amphetamine

Tranylcypromine

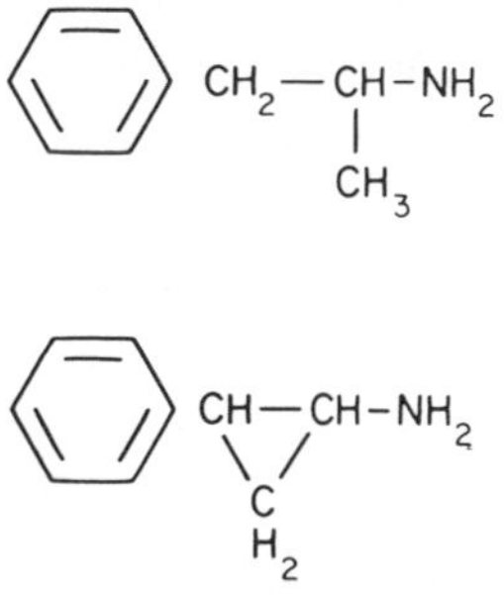

Fig. 3.5 Similarity in structure between
amphetamine and tranylcypromine

and indeed a recent case report on a patient who had taken a tranylcypromine overdose showed that this drug can be metabolized to amphetamine. This is perhaps not too surprising when one considers their structural similarities (Fig. 3.5).

It has been generally assumed that these compounds achieve their therapeutic effect by increasing the function of the monoamines since the major degradative enzyme of these neurotransmitters is inhibited. This has not been seriously questioned; however, the lack of relationship between the onset of enzyme inhibition (in the platelet for example) and clinical response (2–3 weeks) is confusing and may indicate that enzyme inhibition initiates a series of other changes which are finally responsible for the therapeutic effect.

It does appear from clinical data that a high degree of enzyme inhibition is necessary to achieve an antidepressant effect. This view has received recent support from the findings of Mendlewicz and Youdim (1978) where the beneficial effects of a new MAOI (called deprenil) combined with L-5-hydroxytryptophan administration were observed only in patients with at least a 80% inhibition of platelet MAO. However, the trial was not performed 'blind'. Nevertheless this figure of 80% is of interest because it agrees well with animal data, which also suggested that this degree of enzyme inhibition in the brain was necessary to increase monoamine function sufficiently to observe various behavioural changes (Green and Youdim, 1975). It should be noted that administration of MAOIs to patients often does not achieve this degree of enzyme inhibition, as measured by the activity present in human post-mortem brains of people who had been taking MAOIs until death.

Recently there has been renewed interest in the monoamine oxidase inhibitors. This has been predominantly due to the use of deprenil. There are two reasons for this drug receiving such interest. First it does seem to be a potent enzyme inhibitor. Second, it is a selective MAO type B inhibitor (see Section 2.4). Since it is not inhibiting MAO-A in the gut, deprenil administration does not result in the occurrence of the cheese reaction. Furthermore as human (but not rat) brain MAO is predominantly MAO-B, it is a drug that should have good inhibitory effects in the CNS.

Data on the antidepressant action of this drug are scarce at present, but there are several reports of its efficacy when combined with L-dopa in treating Parkinson's disease and this is discussed further in Section 8.15.

3.23   L-Tryptophan plus a MAOI

In 1963 Coppen, Shaw and Farrell reported that L-tryptophan administration potentiated the antidepressant action of a MAO inhibitor and this has been confirmed by several groups of workers. This drug combination undoubtedly causes a marked rise in cerebral 5-HT concentrations and has been quoted in support of the indoleamine-deficiency theory of depression, particularly in view of the failure of L-dopa to potentiate the antidepressant action of a MAOI. Caution should be exercised, however, since this drug combination also increases brain tryptamine concentrations and could thus alter brain catecholamine function by tryptamine formation in catecholamine neurons where it displaces those monoamines. Indeed administration of a MAOI plus L-tryptophan to rats does increase the formation of certain catecholamine metabolites (Eccleston and Nicholaou, 1978) so it cannot be assumed that this drug combination has absolute specificity of action in terms of only altering 5-HT function.

3.24   L-Tryptophan

These considerations lead us in turn to the vexed question of whether L-tryptophan alone is an antidepressant. It has been variously quoted to be as effective as ECT or, conversely, no better than placebo, whilst a recent study placed its efficacy in the order of that seen following imipramine (Jensen *et al.*, 1975). Being an amino acid, and a constituent of normal diet, it is relatively free from side effects and therefore safe.

L-Tryptophan administration increases brain 5-HT turnover because trypto-phan hydroxylase is not saturated with substrate (see Section 2.3). Whether this increased turnover results in increased function, however, is still a matter of debate since it is reasonable to suppose that when MAO activity is not inhibited intraneuronal metabolism of 5-HT by MAO can prevent an increase in transmitter release (see Section 2.5). When tryptophan has been combined with a MAO clinically this intraneuronal metabolism is of course inhibited, which may be why more consistent data have been obtained with the drug combination.

Nevertheless there are enough positive trials to indicate that L-tryptophan can have an antidepressant effect. One suggestion is that its ability to act as an antidepressant is related to the dose given. In support of this contention Young and Sourkes (1977) calculated the mean daily doses given in trials showing an antidepressant action of tryptophan versus those showing no clinical efficacy. The successful trials of tryptophan were found to use a significantly lower daily dose (effective trials dose: 6.1 ± 0.6 g, 6 observations; ineffective trials dose: 8.7 ± 0.9 g, 7 observations; $P < 0.05$).

One way to lower the effective dose would be concomitant administration of a pyrrolase inhibitor. However, despite the claims (on the basis of animal studies) that allopurinol (a drug used to treat gout) and nicotinimide are inhibitors, our own studies on normal volunteers have suggested that neither compound has any marked effect on tryptophan metabolism (Green *et al.*, 1980).

The role of L-tryptophan in the treatment of depression remains uncertain.

3.25 Tricyclic antidepressants

A major method of neurotransmitter inactivation at the nerve ending is by a high affinity, energy requiring, uptake system (Iversen, 1975b) and the major tricyclic antidepressant drugs (Fig. 3.6) inhibit this. As can be seen in Table 3.6

Table 3.6 Inhibition of monoamine uptake by tricyclic antidepressants

Name	IC_{50} (inhibition of uptake) (μM)		
	5-HT	NA	DA
Imipramine	0.50	0.20	8.7
Desmethylimipramine	2.50	0.03	50.0
Chlorimipramine	0.04	0.30	12.0
Amitriptyline	0.49	0.05	4.0
Nortriptyline	1.60	1.30	5.5

IC_{50}, drug concentration to cause a 50% inhibition of [^3H]-mono-amine by brain homogenates *in vitro*. Adapted from Iversen and Mackay (1978) with permission of the British Association for Psychopharmacology.

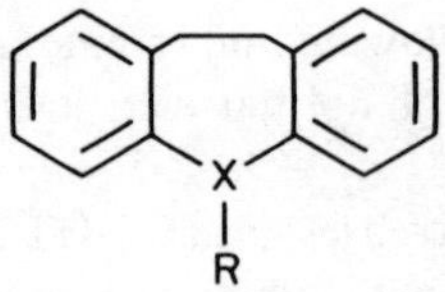

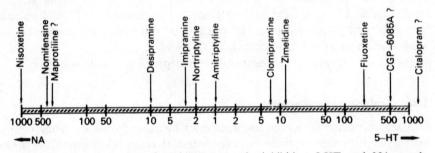

Name	X	R
Imipramine	N	$-(CH_2)_3-N-(CH_3)_2$
Desimipramine	N	$-(CH_2)_3-NH-CH_3$
Chlorimipramine	N (and 3−Cl)	$-(CH_2)_3-NH-CH_3$
Amitriptyline	C	$=CH(CH_2)_2-N-(CH_3)_2$
Nortriptyline	C	$=CH(CH_2)_2-NH-CH_3$

Fig. 3.6 The structure of several common tricyclic antidepressants

Nisoxetine · Nomifensine ? Maprotiline ? · Desipramine · Imipramine Nortriptyline · Amitriptyline · Clomipramine Zimelidine · Fluoxetine · CGP–6085A ? · Citalopram ?

1000 500 · 100 50 · 10 5 · 2 1 2 · 5 10 · 50 100 · 500 1000

◄—NA · 5–HT—►

Fig. 3.7 Relative potencies of antidepressants in inhibiting 5-HT and NA uptake. ? indicates approximate values. (Reproduced from Iversen and Mackay (1979) with permission of the British Association for Psychopharmacology)

they have different potencies in inhibiting the 5-HT, DA, and NA carrier pumps. This is shown in a rather different way by Fig. 3.7. In brain slices imipramine and amitriptyline are more potent than their respective *N*-monomethyl analogues in antagonizing 5-HT, with chlorimipramine being the most potent. Iversen and Mackay (1979) discuss structure–activity relationships in some detail.

There is therefore a degree of selectivity of action of the tricyclics, although in practice not very high, for example chlorimipramine, a drug often used as a specific 5-HT uptake inhibitor, has a major metabolite that is a NA inhibitor.

Recently several new drugs have been produced with greater selectivity. Those displaying a high degree of specificity towards the 5-HT pump are fluoxetine, citalopram, and zimelidine (Fig. 3.8). These compounds are structurally dissimilar

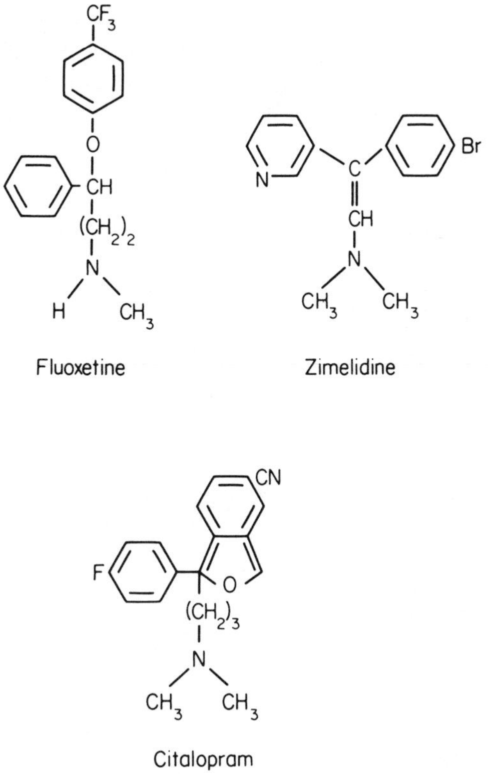

Fig. 3.8 The structures of fluoxetine, zimelidine, and citalopram

from the 'traditional' tricyclics. Similarly some of the new selective NA inhibitors (e.g. viloxazine and nomifensine) are also very different structures from the tricyclics although maprotiline is more closely related (Fig. 3.9). All these compounds show reasonable selectivity (Table 3.7). Nomifensine has an unusual characteristic in that it also displays moderate inhibition of DA reuptake.

It has been generally proposed that tricyclics produce their antidepressant action by increasing the concentration of monoamines in the synaptic cleft by their inhibitory action on the amine uptake pump. The increased synaptic cleft concentration would presumably increase monoamine function, which fits in nicely with the concept of increased monoamine responses being necessary for antidepressant activity. However, as this action was one of the reasons for proposing the theory in the first place this is rather a circular argument! Furthermore it ignores several points and it is amazing that this idea has survived essentially unchallenged for so long.

The first objection is that, in normal animals, tricyclics not only inhibit reuptake, but decrease monoamine synthesis as well. This is almost certainly due to the regulatory feedback system, acting through presynaptic receptors (see

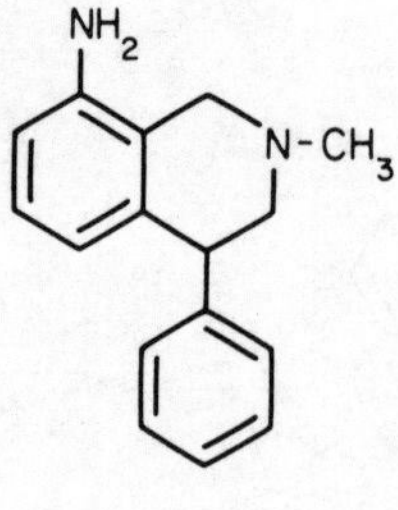

Nomifensine

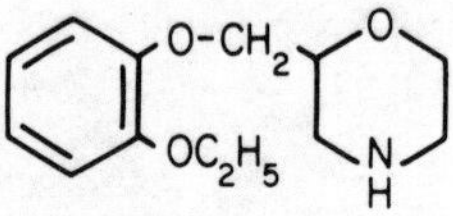

Viloxazine

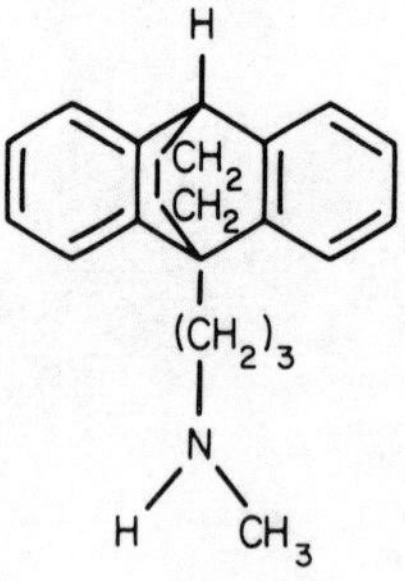

Maprotiline

Fig. 3.9 The structures of nomifensine, viloxazine, and maprotiline

Table 3.7 Selective inhibitors of monoamine uptake

Drug	5-HT uptake	NA uptake	DA uptake
Nomifensine	12.0	0.03	0.14
Maprotiline	30.0	0.08	—
Citalopram	0.014	32.00	—
Nizoxetine	1.00	0.001	0.36
Fluoxetine	0.055	10.00	—
Zimelidine	0.24	2.70	12.0
Mianserin	8.00	0.30	—
Viloxazine	> 100 mg/kg	30.0 mg/kg	—

IC_{50}, drug concentration producing 50 per cent inhibition of uptake, μM. Drugs tested *in vitro* for effects on monoamine uptake by synaptosome preparations from rat brain. From Iversen and Mackay (1978) with permission of The British Association for Psychopharmacology.

Section 2.11). It is of course possible to suggest that such a control mechanism is not occurring in human pathological states; however, there are CSF amine metabolite data to suggest that this 'shut-down' in synthesis does in fact occur in human subjects. This makes it difficult to suggest that these drugs are *increasing* the synaptic monoamine content, when such a good self-regulatory system is coming into play.

The second point is that both the change in monoamine synthesis and blockade of reuptake can be shown to occur in experimental animals very soon after drug administration. In marked contrast tricyclics produce an antidepressant action only after 2–3 weeks administration. The drug enters the brain rapidly and it is therefore not reasonable to suggest that this delay is due to some pharmacokinetic phenomenon, and any theory of possible mechanism of drug action must account for this time scale.

The third point is that cocaine is a moderately potent inhibitor of noradrenaline reuptake at the nerve ending, but does not have an antidepressant action.

The final point is the converse of the cocaine story and has arisen following the development and use of new antidepressant drugs. These compounds seem equivalent in antidepressant action to the tricyclics but do not inhibit monoamine reuptake. Thus antidepressant action and reuptake inhibition do not seem to be inter-related and, assuming for the moment a unitary theory of action, any such theory should incorporate the actions of the newer drugs, which are discussed below.

The first of the newer compounds is iprindole (Fig. 3.10). It has little effect on 5-HT or NA reuptake *in vitro* or *in vivo* following either acute or longer term administration. Another drug with little effect on reuptake is mianserin (Fig. 3.10).

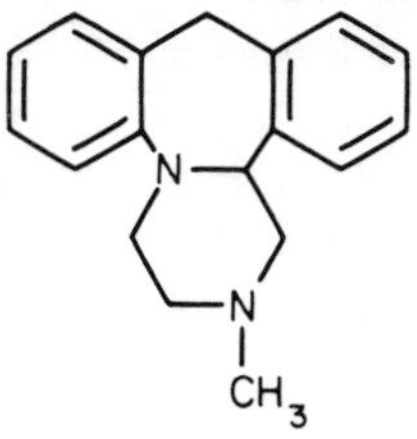

Mianserin

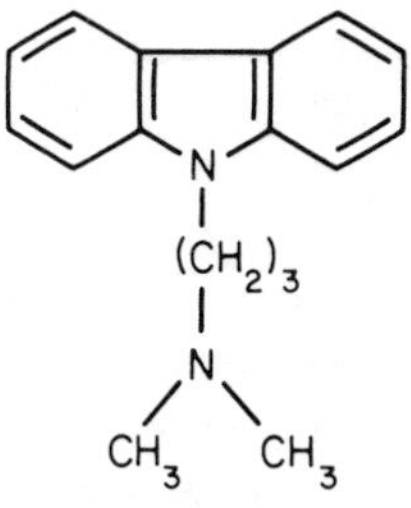

Iprindole

Fig. 3.10 Structures of mianserin and iprindole

This drug was originally developed as a 5-HT antagonist for possible use as a treatment for migraine and its value as an antidepressant was recognized only after it was found to induce human EEG changes similar to those which occur following amitriptyline (Itil, Polvan and Hsu, 1972).

The 5-HT antagonistic action of mianserin is one shared by a variety of the antidepressants as shown by *in vitro* ligand binding studies; however, the effects are often very weak. Baumann and Maitre (1977) showed that mianserin, amitriptyline, and imipramine all increase [^{3}H]-NA release from electrically stimulated cortical slices. They appear therefore to be antagonists of the pre-synaptic α-receptor – the so-called $α_2$-receptor (Langer, 1977). Recently there has been further evidence of this action on presynaptic NA receptors. Crews and Smith (1978) have shown that after longer term treatment (2–3 weeks) with desmethylimipramine the neurotransmitter overflow following electrical stimulation of nerves is enhanced, suggesting a desensitization of the presynaptic α-receptors. Action of the tricyclics at the $α_2$-site has been invoked as one possible mechanism for the therapeutic action of the drugs (U'Pritchard *et al.*, 1978) and it is an attractive theory, particularly as some of the changes are seen only after longer term treatment.

Another theory invoking a similar mechanism of action for a variety of diverse antidepressants is that of Vetulani *et al.* (1976). They have found that chronic but not acute drug administration results in a decrease in the sensitivity of forebrain noradrenaline-sensitive adenylate cyclase to NA. Electroconvulsive shock produced a similar effect when given once daily for a week.

This result has now received support from ligand binding studies (Section 2.14). Bergstrom and Kellar (1979) and others have demonstrated that repeated electroconvulsive shock decreases the number of β-adrenoceptor binding sites. Sellinger-Barnette, Mendels and Frazer (1980) and various other groups have shown that long term administration of a range of antidepressant drugs (tricyclic antidepressants and monoamine oxidase inhibitors) produces a marked lowering in the number of dihydroalprenolol (a β-adrenoceptor antagonist) binding sites. This change is not produced by benzodiazepines or neuroleptics. Nevertheless there is still no clear indication that this change is associated with the antidepressant mechanism of these drugs.

Both these theories therefore incorporate the time lag required for tricyclics to act. The theory of Vetulani, Sulser and colleagues stands existing theories on their heads and proposes that in fact it is NA *overactivity* which occurs in depression and that the drugs and ECT produce a decrease in sensitivity. The data of Baumann and Maitre (1971) on the other hand are less heretical since they suggest that the drugs act by increasing NA function.

It is also worth mentioning the provocative findings of Shopsin *et al.* (1975) that the tryptophan hydroxylase inhibitor *p*-chlorophenylalanine reversed the antidepressant action of imipramine whilst the catecholamine synthesis inhibitor α-methyltyrosine was without effect, pointing to an involvement of 5-HT in the action of imipramine. One would very much like to see both a confirmation of this work and an extension to other antidepressant drugs, although the ethical problems surrounding such a project seem to us to be enormous.

Table 3.8 The potency of various drugs in inhibiting histamine sensitive adenylate cyclase

Drug	K_i (μM)
Chlorimipramine	0.055
Imipramine	0.16
Desimipramine	0.35
Amitriptyline	0.053
Nortriptyline	0.45
Protriptyline	0.48
Iprindole	0.23
Mianserin	0.065
Chlorpromazine	0.041
Haloperidol	0.081
Fluphenazine	0.071

K_i, the concentration of drug required to inhibit the activity of enzyme by 50%. Results adapted from Kanof and Greengard (1978).

A rather different theory of the action of tricyclics has been proposed which does not involve monoamines. Kanof and Greengard (1978) proposed that the drugs might be acting through their antihistaminic actions (Table 3.8). However, the failure of various antihistaminic drugs to act as antidepressives refutes any simple antihistamine hypothesis.

The structure of the side chain of the tricyclics bears a similarity to acetylcholine (Fig. 3.11) and it is therefore not surprising that these drugs are also anticholinergic and this action gives rise to many of the reported side effects. Drugs without such similarity, such as mianserin, have rather fewer of these effects (see the review of Mindham, 1979).

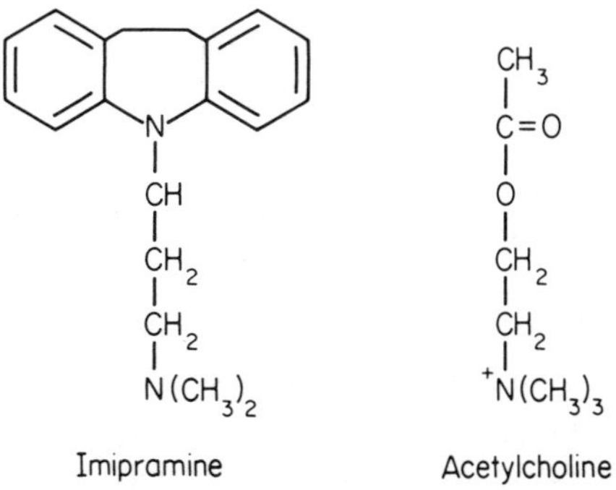

Imipramine   Acetylcholine

Fig. 3.11 Similarity in structure between a tricyclic antidepressant (imipramine) and acetylcholine

3.26   Tricyclic binding to brain and platelet

Recently Langer and colleagues have demonstrated the specific binding of [³H]-imipramine to membranes prepared from rat brain and also to human platelets. This binding has high affinity and a variety of other tricyclic antidepressants (though not atypical compounds such as mianserin) compete for the site (Raisman, Briley and Langer, 1979). Chronic administration of tricyclics was found to decrease the number of [³H]-imipramine binding sites, but not the dissociation constant, in the brain. They also showed that these changes were paralleled by the similar changes taking place in the characteristics of the platelet binding site. Such a result clearly validates studies in which the platelet tricyclic binding characteristics are studied during drug administration in order to understand what is happening in the brain. The important point that arises, however, is the site of the tricyclic binding. Whilst it would be particularly interesting if the tricyclics were binding to a novel site (like the benzodiazepines), work by this group indicates that the binding may be to the 5-HT uptake site. This would then go some way towards explaining their most recent work (Briley *et al.*, 1980) which showed that depressives have lowered [³H]-imipramine binding to platelets, since as reported earlier (Section 3.17) depressives have been reported in several studies to have lowered 5-HT uptake into platelets. The lowered [³H]-imipramine binding did not increase on mood improvement. Whether the platelet is reflecting, in any way, similar changes which are occurring centrally is a matter of speculation at present.

3.27   Dosage of tricyclics

The administration of a standard dose of amitriptyline has been shown to result in an 8-fold range in plasma drug levels in a group of 50 patients and a 10-fold

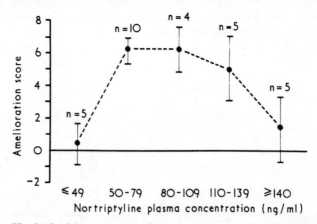

Fig. 3.12   Mean ± s.e.m. of amelioration score for different intervals of plasma concentration of nortriptyline. *n* = number of patients in each group. (Reproduced from Åsberg *et al.* (1971) with permission of the *British Medical Journal*)

range in the plasma concentration of its major metabolite nortriptyline. Coppen *et al.* (1978) have suggested that there is no relationship between plasma level and therapeutic response. However Åsberg *et al.* in 1971 reported a curvilinear relationship between nortriptyline plasma levels and therapeutic effect (Fig. 3.12). This finding has been confirmed several times and has been referred to as 'the therapeutic window'. It is supported by data from Kragh-Sørenson *et al.* (1976) who suggested that in those patients who failed to respond and were found to have a high concentration of drug a reduction in administered dose resulted in clinical improvement.

3.28 Lithium

The role of lithium in the treatment of depression is still uncertain. However, there are now several reports of its efficacy. There have also been some reports suggesting that lithium in combination with a monoamine oxidase inhibitor is effective in intractable cases of depression, resistant to more conventional therapy (e.g. Pare, 1975).

Much has been written about lithium and there are several excellent reviews (Gershon and Shopsin, 1973; Johnson and Johnson, 1978). The problem in writing about its mechanism of action is not that there are no suggestions but rather a surfeit of possibilities.

Studies have usually shown that lithium increases the rate of monoamine synthesis, such changes being seen in both indoleamine and catecholamine systems. It changes monoamine compartmentation and increases intraneuronal metabolism. In experimental animals different effects are seen after different periods of administration and after different doses.

Therapeutically lithium is said to act only when plasma levels are kept within a fairly narrow range. It also has a low therapeutic index, that is to say it has toxic effects at plasma levels not that much higher than therapeutic levels. Fortunately it is a very easy drug to measure routinely in plasma.

At present the use of lithium in depression remains very much a subject for research, the predominant use of this ion being in the prophylaxis of recurrent affective illness (Sections 3.34–3.36).

3.29 Electroconvulsive therapy

Finally we come to electroconvulsive therapy (ECT). ECT remains a major treatment for severe depressive illness and several recent papers have emphasized its safety and efficacy (e.g. Royal College of Psychiatrists, 1977; West, 1981). Nevertheless there have been few suggestions as to how this treatment might work. Recently, however, several pieces of data have indicated a possible mechanism of action.

Initial experiments studied the effects of electroconvulsive shock on various animal behavioural models. These models have been used to assess central

84

monoamine function and rely on the behavioural changes that result from increasing 5-HT or DA synthesis and release in the brain. The behavioural changes were quantified on activity meters. Postsynaptic changes in monoamine function were examined by measuring the behavioural changes following administration of 5-HT or DA agonists.

Following administration of a single electroconvulsive shock (ECS) daily for 10 days rats showed enhanced responses following stimulation of brain 5-HT, or DA systems (see Grahame-Smith, Green and Costain, 1978). Modigh in Sweden obtained data in mice suggesting that repeated ECS also increased NA

Table 3.9   Evidence for altered 5-HT and NA function following repeated ECS

Response examined	Response following ECS
5-HT	
Hyperactivity following administration of tranylcypromine-L-tryptophan	Enhanced
Hyperactivity following administration of 5-methoxy-*N,N*-dimethyl-tryptamine	Enhanced
Hyperactivity following administration of quipazine	Enhanced
NA	
Locomotor response following apomorphine plus clonidine in reserpine pretreated mice	Enhanced
Growth hormone response following apomorphine plus clonidine in reserpine pretreated rats	Enhanced
Sedation following clonidine	Reduced*

For references see Green, 1980. Reproduced from Green (1980) with permission of John Wiley & Sons Ltd.
*Akagi, Green and Heal, 1981.

Table 3.10   Evidence for altered DA function following repeated ECS

Response examined	Response following ECS
Hyperactivity following administration of tranylcypromine/L-dopa	Enhanced
Locomotor activity following administration of methamphetamine	Enhanced
Locomotor activity following administration of apomorphine	Enhanced
Circling behaviour in unilateral nigrostriatal lesioned rats following methamphetamine	Enhanced
Circling behaviour in unilateral nigrostriatal lesioned rats following apomorphine	Enhanced
Locomotor response to dopamine injected into n. accumbens	Enhanced
Locomotor response to dibutyrylcyclic AMP injected into n. accumbens	Enhanced

For references see Green (1980). Reproduced from Green (1980) with permission of John Wiley & Sons Ltd.

postsynaptic behavioural responses, whilst our own recent data indicate decreased presynaptic α_2-adrenoceptor responses (Akagi, Green and Heal, 1981).

Tables 3.9 and 3.10 list the various behavioural changes which are altered by repeated ECS.

As is apparent from this chapter, it is generally suggested that endogenous depression may be treated pharmacologically by increasing, in some way, monoamine function, and tricyclics, MAO inhibitors, and L-tryptophan may all increase the synaptic cleft monoamine concentration. ECT on the other hand appears to increase the postsynaptic monoaminergic response to the same amount of released transmitter.

If the ECS data are at all relevant to ECT then enhanced responses should be seen when it is given in ways closely mimicking ECT administration and responses should not occur when it is given in ways not thought to be therapeutically useful. This hypothesis was tested using the criteria for successful ECT suggested by Fink (1974) and Kety (1974). The data are summarized in Tables 3.11, 3.12, and 3.13 and have been reviewed in a recent chapter (Green, 1980).

Generally a single ECS does not have a marked therapeutic effect: repeated shocks are necessary, generally 6–8 shocks spread over 2–3 weeks (Royal College of Psychiatrists, 1977). A single ECS has no effect on the behavioural models, whereas 10 daily ECS are effective as are 5 ECS spread out over 10 days or 8 ECS spread out over 17 days. In contrast multiple ECT (many shocks in one day) is not generally considered therapeutically useful and multiple ECS has no effect on the animal behaviour.

ECT appears to be effective whether given through unilaterally or bilaterally placed electrodes and whether the current is alternating, unidirectional, or pulsatile (see Fink, 1979). Variations in the current characteristics or electrode placement do not affect the production of the enhanced behavioural responses.

Table 3.11 Conditions necessary for successful ECT and effects of the same conditions on 5-HT induced behavioural enhancement following ECS*

Condition	Effect on 5-HT induced behaviour
Multiple ECS (8 in 1 day)	No enhancement
1 ECS daily for 10 days	Enhancement
5 ECS spread over 10 days	Enhancement
8 ECS spread over 17 days	Enhancement

*Clinical improvement is dependent on the number and frequency of seizures. Clinical change evolves gradually, the change being initiated by ECT and sustained by repeated applications. Multiple ECT in one day is rarely successful.

Reproduced from Green (1980) with permission of John Wiley & Sons Ltd.

Table 3.12 Conditions necessary for successful ECT and effects of the same conditions on 5-HT induced behavioural enhancement following ECS*

Condition	Effect on 5-HT induced behaviour
10 daily flurothyl induced convulsions†	Enhancement (also DA-mediated behaviour)
Bilateral electrode placement sinusoidal current‡	Enhancement
Unilateral electrode placement sinusoidal current‡	Enhancement
Bilateral electrode placement monophasic current‡	Enhancement

*The method of inducing the seizure is less important than the number and frequency of the induced seizures.
†Chemically induced seizures have been used with success.
‡Electrode location, type, and duration of the current does not affect therapeutic outcome (but rather features such as memory loss).
Reproduced from Green (1980) with permission of John Wiley & Sons Ltd.

Table 3.13 Conditions necessary for successful ECT and effects of the same conditions on 5-HT induced behavioural enhancement following ECS*

Condition	Effect on 5-HT induced behaviour
5 ECS in 10 days + neuromuscular blocker	Enhancement
10 daily subconvulsive shocks	No enhancement
10 daily exposures to N_2 (hypoxia)	No enhancement
10 daily ECS to feet	No enhancement

*The persistent therapeutic effects depend on the changes in the CNS which accompany the seizure, not from peripheral components or 'non-specific' effects such as anoxia or 'stress'.
Reproduced from Green (1980) with permission of John Wiley & Sons Ltd.

Neuromuscular blocking drugs, introduced to protect the patient during ECT, do not alter the therapeutic response, suggesting the importance of the central seizure. Administration of fazadinium, a rapidly acting neuromuscular blocking drug, to the rats before the ECS does not prevent the production of the enhanced rat behavioural responses.

The importance of the convulsion is also apparent from the uncontrolled clinical observations that the subconvulsive shocks are ineffective and subconvulsive shocks do not produce behavioural enhancement. Nor does it appear to be important how the convulsion is produced, both photoconvulsions and chemical convulsion (with the inhalant drug flurothyl) having an antidepressant effect (see Fink, 1979). Flurothyl induced convulsions will produce enhanced 5-HT and dopamine-mediated behavioural response.

Finally electroshock to the feet or hypoxia do not have effects on the behavioural models, suggesting that non-specific effects probably do not play a role in the behavioural models.

How then might ECS be altering monoamine function? Data suggest that neither the 5-HT or DA receptor characteristics alter following ECS while the number of β-adrenoceptor binding sites appears to decrease (see Green, 1980).

Perhaps therefore it is changes in other modulatory transmitter systems regulating monoamine responses that are allowing the appearance of enhanced responses. Evidence is accumulating for changes in brain GABA and met-enkephalin biochemistry and function (see Green, 1980) and recently it has been found that destruction of brain noradrenergic systems prevents the appearance of the enhanced 5-HT and DA-mediated behavioural responses (Green and Deakin, 1980).

What the relevance of these behavioural observations is to the therapeutic mechanism of ECT must remain at present a matter of pure speculation, though as stated earlier enhanced monoamine function would fit existing theories well. If the biochemical changes following ECS can be clarified and their relevance to the therapeutic action of ECT established, it gives promise of the development of a pharmacological treatment with the similar high degree of efficacy, but one hopes, greater specificity and fewer side effects.

3.30 General conclusions

Can we put together all the foregoing data on clinical observations and actions of drugs used in depression in any meaningful way?

Theories have been useful in enabling researchers to have a framework on which to work. Here it is worth quoting Professor Hermann Van Praag in his book (Van Praag, 1976):

> 'Quite apart from the fact that it probably contains a kernel of truth, the heuristic value of the monoamine hypothesis for psychiatry cannot be overestimated. It has catalysed an impressive amount of research, ...'

The question therefore is whether the heuristic value is still present. In our view it is. None of the foregoing data on its own is very impressive. Indeed Bunney and Post (1977) have argued convincingly that the pharmacological data do not any longer support the catecholamine hypothesis. Nevertheless, when all observations are taken together, it is hard to resist the conclusion that monoamines are involved in some way in the pathogenesis of depression and that antidepressant action is reflected in mechanisms involving monoamines. There is no reason to suppose, however, that the different treatments have a common mode of action. They may work through several differing mechanisms and this would go some way to explaining why one treatment seems more effective in one patient than in another.

Nor must we assume that the *primary* metabolic or functional abnormality resides in the monoamine systems; it could well be that the initial changes occur in other neuronal systems, and that acting on the monoamines with antidepressants normalizes behaviour without affecting the underlying problems.

The sudden explosive growth of data on the presence of other possible neurotransmitter systems, particularly the peptides, is remarkable.

Some peptides have been examined for antidepressant activity; TRH has been the subject of several trials – mostly disappointing – and there has been at least one trial of β-endorphin.

However, to date much of the experimental data on peptides indicate that these compounds can act to alter central monoamine function, just as there is increasing evidence for close functional relationships between the central neurotransmitters (see for example the book by Garattini, Pujol and Samanin, 1978) which make nonsense of separate theories of catecholamine or indoleamine involvement in depression. Future research is likely to examine monoamine–peptide neuro-transmitter interactions.

PART 4   MANIA

3.31   Psychiatric aspects

As mentioned in Section 3.3 some patients experience episodes of mania alter-nating with their depression. It is also possible to experience manic symptoms without ever getting depressed, but this is uncommon, and patients with recurrent mania usually show symptoms and signs of depression at some stage. It is generally held that treatment of depression especially by MAOIs or ECT may precipitate an episode of mania.

As with the diagnosis of depression the cardinal feature of mania is altered mood, in this case a heightening of mood to cheerfulness, elation, and euphoria. The mood change is more labile than that seen in depression, and whilst some consistent diurnal variations may be observed, more rapid, minute to minute changes are common. Other mood changes such as irritability and hostility are common, as is a general disinhibition.

The other major features are aspects of overactivity; overactive behaviour with constant movement which would in other circumstances be completely exhausting is invariable. Overactive thoughts are evidenced in both the spoken and written word. Flight of ideas, a rapid change in the subject of the thoughts, with increas-ingly tenuous connections between them, punning and rhyming, together with a general pressure of talk, make any effective dialogue difficult.

In the severer forms of mania delusions and hallucinations may occur with sometimes ideas of reference and passivity feelings (see Section 5.1).

Although manic patients usually feel very well, have little insight into their con-dition and may even generate a good mood in those around them, the elation results in a loss of rational decision making and commonly financial difficulties due to overspending, or legal difficulties following lack of the usual constraints on behaviour, are a consequence.

Historically mania was subtyped into several classes, hypomania, delusional mania, and delirious mania, but it is likely that this is an expression largely of degree. The distinction between hypomania and mania in particular has been obscured by modern drug treatment and most patients now seen would probably have been classified as hypomanic.

3.32 Biochemistry

Relatively little work has been done on the biochemical changes that occur in mania. This may, in part, be due to the difficulty in obtaining drug free patients but also the relative scarcity of patients suffering only from mania.

Clearly in those patients with manic depressive episodes there is the added complication of assigning any change seen specifically to a role in mania, or indeed depression.

There are several reports that lumbar CSF 5-HIAA concentrations are decreased in mania; however, others have reported both normal and raised 5-HIAA concentrations (see Van Praag, 1976).

Data on lumbar CSF HVA concentrations are also conflicting, although there are several reports of lowered concentrations and MOPEG has also been reported to be both low and normal.

Using probenecid accumulation of metabolites it has been shown that manic patients have lowered turnover of both 5-HT and dopamine.

There also seems to be indications for abnormal noradrenergic metabolism. The general consensus from at least 10 studies is that urinary MOPEG excretion is increased during mania (see Schildkraut et al., 1977). However, the available data on CSF MOPEG do not support the view of an increase in mania.

Overall one may conclude that there is certainly no evidence for either the 5-HT or catecholamine systems being changed in 'opposite' directions in depression and mania, and, if anything, biochemically these illnesses show distinct similarities.

Second, the data on HVA do not support the proposal that HVA concentrations are lowered in depression because of motor retardation (Section 3.18).

3.33 Pharmacology – introduction

In general the initial treatment of mania is the use of a major tranquillizer such as haloperidol. This has the effect of making the patient 'manageable'. Lithium (always in the form of a salt, usually the carbonate) also has an antimanic action but usually takes over a week to work and is thus not usually the drug of choice on admission to hospital.

The antimanic action of the major tranquillizers is presumably through their dopamine antagonist actions (Sections 5.15–5.18).

Whilst lithium does not act rapidly as an antimanic drug, its action as a prophylactic drug to prevent further manic episodes is proven. Whether it does so by preventing the illness or treating the underlying illness when it occurs has not been ascertained.

3.34 Lithium – historical aspects

Lithium was used during the nineteenth and early twentieth centuries as a sedative in the form of lithium bromide and although it was claimed to be the most efficacious bromide salt, it is now impossible to know whether this claim was true.

The use of lithium (as the chloride) reappeared during the 1940s as a salt (NaCl) substitute in food for hypertensive patients but rapidly lost favour as it caused severe toxicity and several deaths.

Lithium was first used in psychiatric disorder by John Cade in Australia. He was examining whether manic patients excreted a toxic metabolite and found that urea produced hyperexcitability in guinea pigs and noted that lithium urate protected against this excitation. Since the 'control' substance, lithium chloride, had the same effect, he concluded that it was lithium which was the beneficial agent. Whilst it has now been conjectured that part of the effect might have been toxicity, the impact of these experiments on modern treatment of mania should not be underestimated. This observation stimulated Cade to try lithium therapy in 10 manic patients and the improvement was dramatic. However, the report of Cade (1949) was almost unnoticed, although there were confirmatory reports, and general acceptance was slow.

The reasons for this are not hard to find. First, lithium was a compound that had achieved a notorious reputation following the toxicity caused in the hypertensive patients using it as a salt substitute. Second, being a natural ion and therefore cheap and not patentable, the drug industry was not interested. Third, there was scepticism that a mineral could have a specific effect.

The credit for the popularization of lithium is mainly due to Mögens Schou in Denmark who carried out many carefully controlled trials which showed the value of lithium, and to Gershon in the USA who also made extensive investigations of this compound.

The value of lithium in the prophylaxis of mania is now accepted. There are several good reviews on the biochemistry and pharmacology of lithium (Johnson and Johnson, 1978; Gershon and Shopsin, 1973).

3.35 Measurement and therapeutic plasma levels of lithium

Lithium is straightforward to measure in blood using atomic absorption spectrometry or flame photometry. It is measured regularly as it is generally effective between well defined concentrations (0.8–1.2 mEq/l serum). Lower levels are ineffective whilst only slightly higher concentrations (greater than 2.5 mEq/l) result in toxicity, including nausea, muscular weakness, fine tremor, polydipsia, polyuria, oedema, and at higher levels, tremor, vomiting, diarrhoea, confusion, and death.

In addition the function of the thyroid gland needs to be checked regularly as chronic lithium treatment carries a risk of causing hypothyroidism.

3.36   Pharmacology of lithium

The mechanism of the antimanic (and antidepressant) action of lithium is unclear. The problems in trying to define a possible 'mode of action' for lithium are compounded by the fact that there have been marked variations in the dose, route of administration, and duration of treatment of lithium in the various studies. Several investigations have clearly shown that the effects of lithium administration on (for example) 5-HT metabolism change markedly during the period of administration.

Partly as the result of Schildkraut's catecholamine hypothesis of affective disorders and the fact that there is some evidence for abnormalities of catecholamine metabolism in mania, there has been much work on the effect of lithium on catecholamines.

Lithium given for a few days has been shown to increase the rate of intraneuronal metabolism of NA and increase NA turnover. However, following longer term administration (4 weeks) there is little evidence for an alteration in brain NA or DA concentrations or rates of turnover (see Schildkraut, 1973).

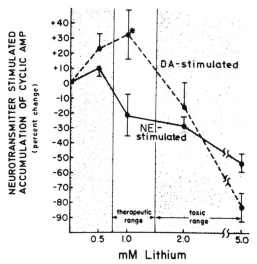

Fig. 3.13   The effect of lithium on NA- and DA-sensitive adenylate cyclase activity (measured as the stimulated accumulation of cyclic AMP in guinea pig brain. DA-stimulation in caudate nucleus (– – – –); NA-sensitive in cortex (——). Different from control $P < 0.001$. (Reproduced from Ebstein, Eliashar and Belmaker (1980) with permission of John Wiley & Sons Ltd)

Recently Ebstein, Eliashar and Belmaker (1980) have shown in rats that at therapeutic plasma concentrations of lithium there is a marked inhibition of NA-stimulated adenylate cyclase, while DA-stimulated adenylate cyclase activity shows an increase (Fig. 3.13), confirming earlier reports. It is interesting that inhibition of NA-sensitive adenylate cyclase has been reported following administration of a range of antidepressant drugs (Section 3.25). Whether these changes have relevance to the therapeutic action of lithium remains unclear.

There are several reports that short term (3–6 days) lithium administration increases 5-HT turnover and there are data which suggest that it alters the compartmentation of 5-HT at the nerve ending.

Whilst, therefore, there is good evidence for lithium altering monoamine metabolism, it is still unclear whether these changes have any importance in the clinical action of the drug.

3.37   Other pharmacological approaches

Other drugs that received interest in the treatment of mania have included methysergide, a 5-HT antagonist. Various trials resulted, initially, from the hypothesis that there was 5-HT overactivity in mania. However, after a couple of early positive results further carefully controlled trials failed to find any beneficial effect and one report suggested it made some patients worse.

It was also reported that α-methyl-p-tyrosine had an antimanic effect but the toxicity of this compound precludes its clinical use although it has value in indicating the possible involvement of catecholamines in mania.

Chapter 4

Anxiety

PART 1 PSYCHIATRIC ASPECTS

4.1 Symptomatology

Anxiety, even more than depression, is experienced very commonly. Not only is it common but it can clearly be seen to have beneficial effects in ordinary behaviour. This has been described by Yerkes and Dodson in 1908, who showed an inverted U-shaped relationship between performance and anxiety, which has since become a well established 'law' (see Fig. 4.1).

At low levels of anxiety, behaviour may not be purposefully directed and performance low. There is a middle range of anxiety where optimal behaviour has been reached. Above a certain level, a further increase in anxiety interferes with performance.

There are clearly difficulties in deciding when anxiety is abnormal, but the following three points have proved to be useful guidelines.

1. The experience of anxiety should be perceived as unpleasant by the patient and not as being simply 'keyed up' in an appropriate way.

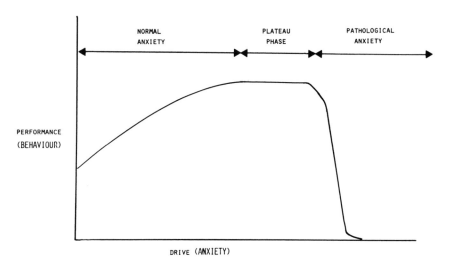

Fig. 4.1   The relationship between anxiety and performance

93

2. The anxiety should not be under voluntary control; that is that the patient is unable to distract himself from it, either by a mental process, or by, for instance, working harder, and is not easily distracted by others.
3. The anxiety is out of proportion to the subjects being worried about. This condition may be inappropriate if the patient has severe problems such as imminent financial disaster or marital collapse.

Although the above criteria are to some extent susceptible to subjective evaluation, they do provide a fairly reliable yardstick by which to judge anxiety.

The psychic experience of anxiety is experienced and described in a variety of ways. Fear, panic, apprehension, foreboding, these feelings are accompanied to variable extents by the somatic or physical components of anxiety. These sensations are the consequences of sympathetic nervous system overactivity, and many may be reproduced by injections of noradrenaline.

Signs and symptoms are seen in several systems:

1. cardiovascular: tachycardia or increased heart rate often experienced as palpitations, increased blood pressure;
2. muscular: increased tension of voluntary muscles produces muscular tremor, or pain, for example headaches or back pain;
3. gastrointestinal: dry mouth due to decreased salivary secretion, feelings of a lump in the throat (muscle tension), nausea and 'sinking' feeling in the stomach, diarrhoea;
4. respiratory: a subjectively experienced difficulty in breathing which may make the patient breathe more rapidly and may result in dizziness and temporary loss of consciousness;
5. sweating: increased sweat production.

Whilst the subjective experience of anxiety is difficult to quantify, other than by asking the subject to say how anxious he feels, or to rate himself on a simple scale, the physical symptoms may be directly measured, and thus may give some objective rating of anxiety. The correlation between these and subjective reporting is unfortunately sometimes low.

The measures most often used are:

heart rate,
blood pressure,
muscle tension,
skin electrical conductance (which lowers as the skin moistens during sweating).

These changes may be demonstrated to the patient, and have been successfully used with psychological approaches to anxiety where the patient is taught to modify his own anxious responses.

Anxiety may be a symptom in all psychiatric disorders, and is usually regarded as being a secondary feature, although in depression it may be difficult to decide which predominates. In the absence of other primary symptoms, however, it assumes the status of a diagnosis – anxiety state.

4.2 Classification

The classification of anxiety states is not as well investigated as some other psychiatric syndromes, possibly because the main differentiation is related to the causes of the anxiety or to those circumstances in which it arises. Classification has been related to this rather than to differentiating between different types of anxiety symptoms.

An exception to this is the syndrome of 'panic attacks' where psychological and physical symptoms of anxiety are experienced unpredictably, acutely, and severely, and tend to be self-limiting, lasting only for a few minutes.

Some anxiety states are clearly situational, for example agoraphobia (fear of open spaces), or fear of spiders, or otherwise directed such as anxiety over health or other personal problems, and this object can form the basis for classification. In some patients, however, anxiety appears not to be related to specific situations or ideas, and is described as free floating.

A suggested scheme for considering anxiety is as follows.

1. Anxiety secondary to other psychiatric illness
2. Primary anxiety states
   (a) Anxiety neurosis
   (b) Phobic anxiety
       (i) Social phobia (e.g. meeting people)
       (ii) General phobia (e.g. agoraphobia)
       (iii) Specific phobia (e.g. spider phobia)

Although some psychiatrists argue that most if not all anxiety is related to particular situations there is some evidence that 1 and 2 above are distinct groups.

First, genetic influences appear to play a part in the aetiology of anxiety neurosis but not phobic disorders as judged by the incidences in monozygotic and dizygotic twins. Second, there seems to be a differential response to treatment in that anxiety neuroses respond well to drugs and poorly to psychological or behavioural treatments, whereas the reverse is true for phobic states.

4.3 Obsessional disorders

It is convenient to consider these with anxiety states as many of the clinical symptoms arise out of behaviours undertaken in order to avoid anxiety.

The essential components of these is the obsessional thought or rumination, the content of which is partly determined by social and personal factors. Common are thoughts of contamination, especially faecal, or thoughts of having harmed others in some way, or of not having completed simple tasks such as switching off lights, cookers etc., or locking doors. The thoughts give rise to anxiety which may be alleviated by repeated washing, checking etc. or by invoking some magical ritual, for instance counting or doing things a fixed number of times. Commonly there is resistance to these repeated acts or thoughts.

The incidence of these disorders is very low and their aetiology obscure. Likewise there are a few data on their biochemistry or pharmacology. Claims that some drugs (e.g. clomipramine) have a specific action, have not been substantiated and behavioural approaches remain the treatment of choice. However, symptoms of depression are common, and where these respond to antidepressant medication, secondary improvement in the obsessional disorder may be seen.

PART 2   PHARMACOLOGY

4.4   General introduction

Essentially nothing is known of clinical biochemical changes associated with anxiety states apart from a few studies on plasma catecholamine concentrations which found that the concentration correlated with anxiety ratings (e.g. Wyatt *et al.*, 1971). Nevertheless there are safe and effective treatments for anxiety and the mechanisms by which these drugs work are now being elucidated.

There are several classes of drugs that are, or have been, used to treat anxiety. However, the benzodiazepines, now by far outsell all rivals, having an equal efficacy but much greater safety than other drugs, particularly the barbiturates.

4.5   The benzodiazepines

The first benzodiazepine introduced clinically was chlordiazepoxide. This rapidly became the drug of choice in treating a variety of anxiety states. Undoubted reasons for this were the low toxicity of the compound coupled with its relative freedom from producing an addictive state. Psychiatric patients may well use prescribed drugs for self-poisoning, and drugs with the above properties are highly desirable. The relatively low toxicity of chlordiazepoxide was therefore of great value. Furthermore it did not have the respiratory depressant and hypotensive effects of barbiturates. In addition benzodiazepines do not induce the drug metabolizing enzymes (again in contrast to barbiturates), thereby decreasing the risk of drug interactions with other compounds.

Following the introduction of chlordiazepoxide, the pharmaceutical company Roche released diazepam and nitrazepam (see Fig. 4.2). Chlordiazepoxide has decreased markedly in popularity and diazepam is now probably the most widely used benzodiazepine, having a similar pharmacological profile to chlordiazepoxide, but greater potency. Currently there is a bewildering number of 1,4-benzodiazepines on the market (some shown in Fig. 4.2), but there is little reason to suppose that there are marked pharmacodynamic differences between these drugs, despite some claims to the contrary. (However, see Section 4.9 on the metabolism of benzodiazepines.)

The benzodiazepines have three major clinical uses, as anti-anxiety agents, anticonvulsant drugs, and hypnotic (sleep inducing) drugs. Whether a common mechanism underlies these diverse actions is controversial.

In experimental animals the anti-anxiety effect is best illustrated by the conflict punishment model and it is this model which is used to screen potential drugs.

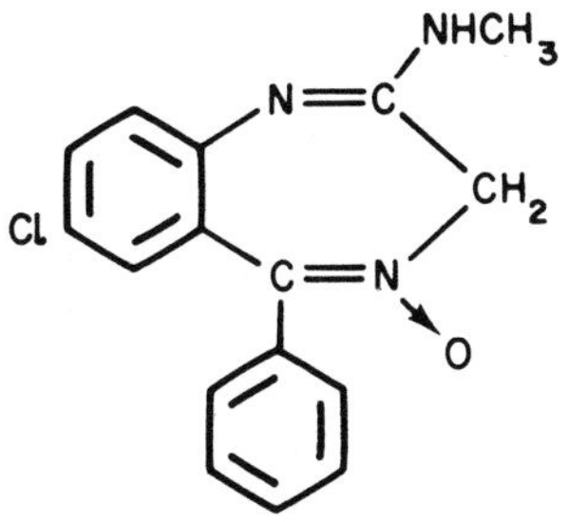

Chlordiazepoxide

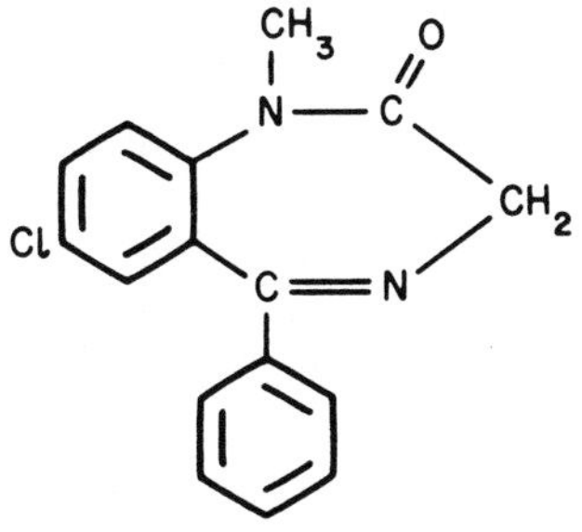

Diazepam

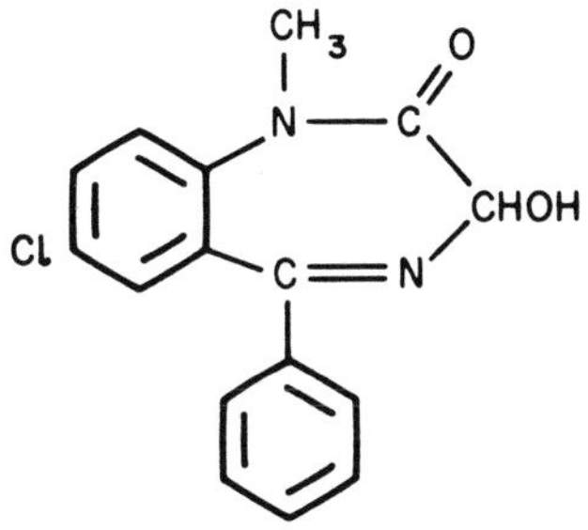

Temazepam

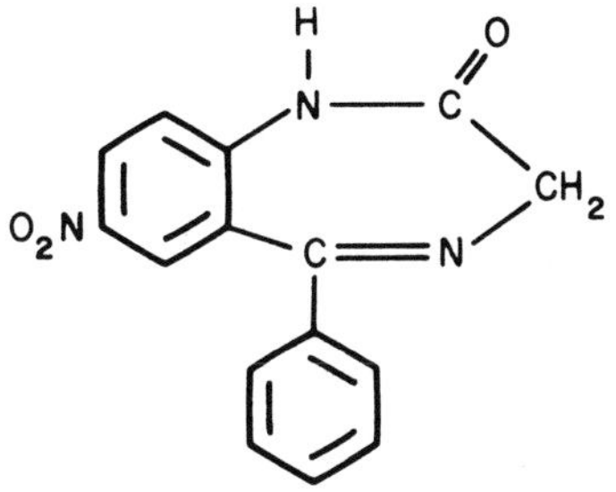

Nitrazepam

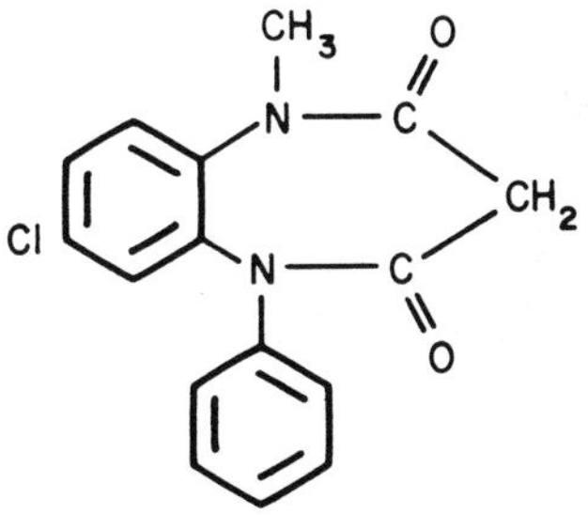

Clobazam

Fig. 4.2 The structures of several benzodiazepines. Chlordiazepoxide, diazepam, temazepam, and nitrazepam are 1,4-benzodiazepines and clobazam is a 1,5-benzodiazepine

Rats are trained to push a lever following a signal, such as a light coming on, to obtain a reward. When the animals are fully trained and always push the lever following the light signal, the system is changed. The light is switched on but when the rats push the lever they now, as well as getting the reward also receive an aversive stimulus, such as an electric foot shock. Not surprisingly (in anthropomorphic terms) they stop pushing the lever. However, when administered benzodiazepines, this suppression of the learned behaviour is abolished. Thus despite knowing that they will get both a reward and a punishment, the latter no longer seems to 'worry' them. This test separates benzodiazepines from many other psychoactive compounds and it can be demonstrated that it is not an analgesic or sedative effect.

There have been some indications that benzodiazepines act in this model by altering 5-HT function (Stein, Wise and Belluzi, 1975). The effect of benzodiazepines in restoring the suppressed responding in punished animals can be mimicked by administration of p-chlorophenylalanine, the tryptophan hydroxylase inhibitor, or by lesions of the brain stem raphe system, thereby producing profound depletion of brain 5-HT concentrations (Tye, Iversen and Green, 1979).

However, while these data suggest that benzodiazepines may act through 5-HT systems, it has also been shown that the restoration of punished responding by 1,4-benzodiazepines can be blocked by picrotoxin and enhanced by facilitation of GABA transmission, suggesting an involvement of GABAergic systems (Costa, 1979, and see Section 4.8). The possibility that these drugs alter GABA function would certainly be in accord with their anticonvulsant action (Section 4.11). Current research has now strengthened the view that benzodiazepines interact with brain GABAergic systems and in turn has provided indications for the following novel mechanism for controlling the function of this inhibitory neurotransmitter.

4.6 Brain benzodiazepine receptors and GABA modulin

Toffano, Guidotti and Costa (1978) have demonstrated that the brain contains a protein of fairly high molecular weight ($\simeq$15,000) which interacts with GABA receptor sites to inhibit their ability to bind GABA. This compound can therefore markedly alter GABA function.

The year before the demonstration of GABA modulin, Squires and Braestrup (1977) had shown that radiolabelled 1,4-benzodiazepines bind with high affinity to specific sites in the brain. This finding caused much surprise since it suggested that these drugs might be interacting with a site normally occupied by an endogenous neurotransmitter or neuromodulator (Iversen, 1977).

A variety of pharmacological and electrophysiological investigations had already indicated that 1,4-benzodiazepines facilitated GABA neurotransmission (see review of Guidotti, 1978). In consequence, Guidotti, Toffano and Costa (1978) suggested that benzodiazepines were binding at the site occupied by GABA modulin. Occupancy of the GABA modulin site by the benzodiazepine, acting as an antagonist at the site would prevent the presence on the site of GABA modulin. The result would be the observed facilitation of GABA transmission. In support of this contention was the fact that benzodiazepines do not interact with the GABA receptor but appear to act postsynaptically (see Costa, 1979). Further, there is a reasonable relationship between the potency of various benzodiazepines as anxiolytics and their potency in competing with GABA modulin.

The potential importance of this finding on our views on the regulation of central neurotransmitters cannot be overemphasized. Hökfelt and colleagues have demonstrated the possible coexistence of several peptides with monoamine neurotransmitters (see Section 2.26). It is indeed possible that the function of several transmitters is regulated by peptides or proteins. Until now central transmitter function in psychiatric patients has been altered by drugs changing the synthesis of the transmitters or blocking the postsynaptic receptor site, with all the attendant problems associated with long term administration of such drugs, such as tardive dyskinesias (Section 5.19). If the function of transmitters other than GABA can be altered by drugs acting on the appropriate modulatory factor then a new era of psychopharmacological drug development could emerge.

4.7   Is there an endogenous benzodiazepine?

The observation that opiates bind to specific receptors in the brain led to the suggestion that the brain might produce an endogenous opiate acting on these receptors and subsequently to the isolation and characterization of the enkephalins (Section 2.28). Similarly the existence of the benzodiazepine receptor has led to several investigations into the existence of an endogenous ligand (other than GABA modulin) for the receptor.

Two suggested ligands have been nicotinamide and inosine. However, both compounds bind only weakly to the receptor *in vitro* and very high doses are necessary to produce a modest rise in seizure threshold.

Recently Braestrup, Nielsen and Olsen (1980) isolated from human urine and rat brain, ethyl β-carboline-3-carboxylate (β-CCE; Fig. 4.3). This compound binds potently to the benzodiazepine receptor. It is not itself the endogenous ligand as the esterification of the compound occurs during the isolation procedure; however, Braestrup has suggested that it is closely related to it. Furthermore this compound appears to have some ability to bind differentially to different classes or types of benzodiazepine receptors.

Interestingly, however, initial pharmacological studies of β-CCE have revealed that unlike the benzodiazepines which raise seizure threshold, it actually markedly lowers it (Cowen *et al.*, 1981; Fig. 4.4). This raises the possibility that the endogenous compound, if it is closely related to β-CCE is a convulsant and anxiogenic compound and that benzodiazepines are antagonists at the β-carboline receptor site.

β-carbolines are compounds known to exist in plants. Tryptophan and tryptamines condense with aldehydes to produce β-carbolines (Fig. 9.2) and this reaction can be shown to occur *in vitro* in CNS homogenate preparations. It has

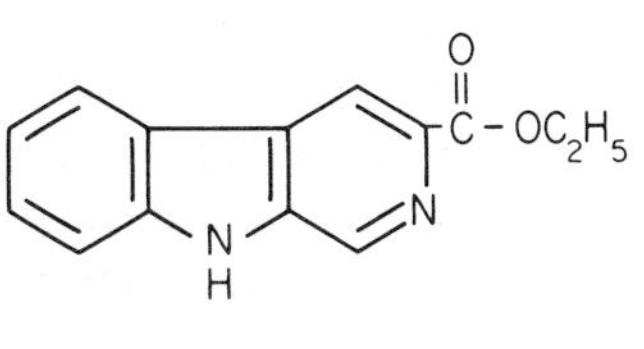

Fig. 4.3   Structure of ethyl β-carboline carboxylate (β-CCE)

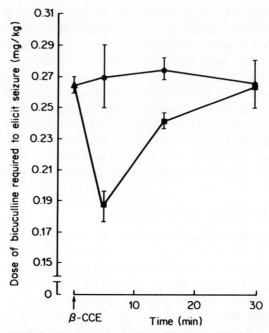

Fig. 4.4 The effect of ethyl β-carboline carboxylate (β-CCE) on seizure threshold. Rats were injected with β-CCE (5 mg/kg) or the drug vehicle intravenously and the intravenous dose of bicuculline required to elicit a seizure at various times following the β-CCE injection measured. ▲, untreated; ●, vehicle injected; ■, β-CCE injected. Lowering of seizure threshold statistically significant at 5 and 15 min after injection

been suggested that they may have a neuroregulator role (Elliot and Hollman, 1977) and they have recently implicated in alcoholism (see Section 9.6).

4.8 Is the anxiolytic action associated with GABA?

The suggestion that the association between benzodiazepines and GABA explains the anticonvulsant action is not unreasonable in view of the fairly long held view that GABAergic systems are involved in seizure activity. Decreasing brain GABA concentrations or blocking receptors with bicuculline produces convulsions and experimental seizures can be blocked by increasing brain GABA concentrations.

In contrast the association of GABA with anxiolytic action is less certain. Nevertheless the intimate association of GABA with many other neurotransmitter systems means that it is not unreasonable to suggest that alteration of this transmitter could alter anxiety states. It could also be that the apparent involvement of 5-HT in the anxiolytic action results from GABA-5-HT neurotransmitter functional interactions. Data are now being published which support this hypothesis. Soubrié *et al* (1981) have provided biochemical evidence for an inhibition by GABA of 5-HT release in the substantia nigra, and Forchetti and Meek (1981) have shown that GABA exerts a tonic inhibitory control on 5-HT neurones in the median raphé nucleus and that benzodiazepines potentiate this action. It is possible therefore that the action of the benzodiazepines in restoring punished responding is through a decrease of 5-HT function, thereby explaining why 5-HT synthesis inhibition has the same effect.

4.9 Benzodiazepine metabolism and kinetics

One major factor which must be borne in mind when considering the pharmacological actions of different benzodiazepines is that many of them are metabolized to the same active metabolites. Thus chlordiazepoxide, diazepam and medazepam are all metabolized through desmethyldiazepam (itself active) and with temazepam are metabolized to oxazepam; a compound which is itself marketed (Fig. 4.5). Tyrer (1974) has pointed out that there is little justification for the marketing of so many compounds.

In general, the benzodiazepines have a long half-life and the half-life of desmethyldiazepam is variable but may be as long as 90 hours (Hillestad, Hansen and Melsom, 1974). The accumulation of this and other metabolites over time makes one question the use of such benzodiazepines as hypnotics, as appreciable quantities of the drugs will be present during the daytime. For this reason the drugs of choice as sleep inducing agents are those benzodiazepines with short half-lives, and either no active metabolites or metabolites with short half-lives, for example oxazepam or temazepam.

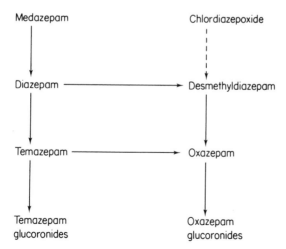

Fig. 4.5 The metabolic relationships between some benzodiazepines. Dotted line indicates metabolism via intermediates

4.10 Dependence on benzodiazepines

Whilst the benzodiazepines are undoubtedly very safe drugs, evidence on their 'addictive' qualities has recently been documented. This is discussed in Section 9.20.

4.11 Benzodiazepines and seizure disorders

While this book is not concerned with seizure disorders it should nevertheless be mentioned that there is also a good correlation between clinical anticonvulsant activity and *in vitro* benzodiazepine binding potency in the brain (Squires and Braestrup, 1977). Furthermore *in vivo* binding and seizure protection show a good correlation (Paul *et al.*, 1979).

4.12 Barbiturates

Barbiturates used to be prescribed as anxiolytics. However, their use has been superseded by the safer and more effective benzodiazepines and these drugs have no place in the modern treatment of anxiety. A major problem with the barbiturates was that of drug dependence. This problem continues today with illicit use and is discussed in Sections 9.21–9.24.

4.13 β-Adrenoceptor antagonists

It may seem odd that the 'beta-blockers' are included in the section on anti-anxiety drugs. In the USA these drugs have not been approved by the Food and Drug Administration (FDA) for use as anti-anxiety drugs. However, in the UK there have been a variety of reports on their use, the first nearly 15 years ago (Granville-Grossman and Turner, 1966). At the doses used it is probable that these drugs are acting via a peripheral β-blockade as there appear to be no central effects (see, for example, Lader and Tyrer, 1972) and practolol, a β-blocker which does not enter the brain, is also effective (Bonn, Turner and Hicks, 1972). These drugs relieve predominantly the physical manifestations of anxiety rather than the subjective experience, which fits in with the suggestion that they act peripherally.

4.14 Meprobamate

This compound was proposed as an anxiolytic drug with a safer, more selective, profile than the barbiturates. In fact, it seems to lie somewhere between the barbiturates and the benzodiazepines in terms of specificity of anxiolytic action, since it does have some sedative action. Like the barbiturates, meprobamate has been largely superseded by the benzodiazepines.

4.15 Tricyclic antidepressants

Some tricyclic antidepressants (see Section 3.25) have distinct anxiolytic properties and are thus used in depressed patients who also have anxiety symptoms. However, there are no indications for the specific use of tricyclics in anxiety, although they may relieve anxiety secondarily to their antidepressant action.

4.16   Monoamine oxidase inhibitors

It has been suggested that the MAO inhibitors are of use in anxiety. However, the problems of dietary induced interactions (Section 3.22) and the lack of specific indications for their use makes their place in the treatment of anxiety uncertain.

Chapter 5

Schizophrenia

PART 1   PSYCHIATRIC ASPECTS OF SCHIZOPHRENIA

5.1   Introduction

Schizophrenia is probably the most problematic disorder in psychiatry. Twenty per cent of British National Health Service hospital inpatients have this diagnosis, reflecting both the prevalence, and the severity, in the sense of long term outcome. Its importance in psychiatric practice derives largely from this, but also from the severe and troublesome nature of the symptoms.

Schizophrenia patients are the paradigm for the lay concept of madness or insanity, but there is much confusion and misconception surrounding the nature of the condition. Until recently there have also been great differences of opinion amongst psychiatrists as to how to diagnose schizophrenia. Such difficulties do not, however, invalidate the concept, which remains essentially the same as when the term was introduced by Eugen Bleuler in 1911 – that of a splitting of psychic functions as evidenced by loosening of association of ideas, detachment from reality, and inappropriate emotions and responses. Bleuler's concept was an attempt to tighten up the concept of 'dementia praecox' outlined by Kraepelin in 1919, by considering only the symptoms he considered specific. Unfortunately these are difficult to define and he had the opposite effect, that of allowing a wider interpretation of the diagnosis. It will be seen in Section 5.5 that modern diagnostic criteria employ a wider range of symptoms. It is worth emphasizing that the concept does not refer to any duality of personality or behaviour.

5.2   Clinical features

The difficulty central to the diagnosis of schizophrenia is that it is made on the basis of symptoms and signs, and there is as yet no objective way of validating a diagnosis so made. Since the diagnosis probably includes several different illnesses presenting with similar symptoms, the presence of a criterion validating the diagnosis as a whole is unlikely, but subgroups with specific biochemical or aetiological associations may be found. None of the features of schizophrenia is specific to it, but to some of them, when they occur in the absence of confusion or disorientation (evidence of organic brain disease), are given diagnostic importance. These were categorized by Schneider (1956) and are known as 'first rank' symptoms. They are as follows.

1. *Delusional perception*: a fully formed delusion which arises from a normal perception.
2. *Passivity experiences*
    (a) The experience of an external force putting thoughts into the patient's mind (thought insertion) or taking them away (thought withdrawal).
    (b) The experience that the patient's thoughts are shared by others (thought broadcasting).
    (c) The experience of sensations, feelings, impulses, or acts being under external control.
3. *Auditory hallucinations*
    (a) Voices repeating the patient's thoughts out loud, or anticipating them.
    (b) Two or more hallucinatory voices discussing or arguing about the patient, referring to him in the third person.
    (c) Voices commentating on the patient's thoughts or behaviour.

Whilst characteristic, these symptoms are not the only ones occurring in schizophrenia.

It is convenient to consider the symptoms under four headings.

1. *Thought disorder*: examples of this are 1 and 2 above. This may be sub-divided into four common components.
    (a) disorders of form; a disordered way of thinking; where the organization or continuity of thought is impaired.
    (b) disorders of content; the presence of irrational beliefs or delusions, commonly of persecution.
    (c) disorders of the stream, where the train of thought is impaired. Thought blocking, the sudden cessation of a train of thought is the most common.
    (d) disorders of possession: the passivity feelings described above.
2. *Perceptual disorders* or hallucinations including auditory hallucinations in 3 above. Other hallucinations may occur: of strange smells or tastes, tactile sensations, or more rarely visual hallucinations.
3. *Emotional disturbances*: emotions are often lacking – 'flattening of affect' – particularly in the later stages of the illness. Motivation is also frequently impaired leading to a general negativism. These features are of importance in determining outcome (Section 5.6).
4. *Motor behaviour*: activity may change either towards a lowering, which may amount to a stuperose state or towards overexcitement and violent over-activity (catatonia). Drug effects on motor performance are complex and may make assessment difficult. The main effect of the drugs most commonly used, the phenothiazines, is that of lowering motor activity, but a number of different involuntary motor behaviours may be produced and possibly confused with some schizophrenic stereotypes.

In addition to the above symptoms of schizophrenia, changes in social functioning and interpersonal relationships, are a long term consequence of the condition, with patients failing to return to their premorbid role in life and showing a tendency to drift down the socioeconomic scale. These 'negative' features may be the most prominent in later stages of the illness, and when they occur in the absence of positive symptoms (delusions and hallucinations) may lead to a diagnosis of 'simple schizophrenia'. Diagnosis solely on this basis is regarded by many psychiatrists as fundamentally unsound, but the concept survives, albeit rarely used.

Other forms of schizophrenia have also been delineated, hebephrenic, catatonic, and paranoid. Such distinctions are rarely made now, and are thought to represent differences in symptomatology which are determined by age of onset and social factors.

5.3 Schizo-affective disorders

These patients have the characteristics of schizophrenic patients but also have significant affective symptoms, of either depression or mania. It is not reasonable to subsume either diagnosis under the other, and they are usually regarded as a separate entity. Although there is some evidence that the response to treatment is more similar to schizophrenia than to depression or mania, the long term outcome may be better than for schizophrenia and for research purposes it is desirable to regard them as a distinct group.

5.4 Diagnosis

It has been mentioned above that there has been dispute over the diagnosis of schizophrenia. This has essentially been because of a difference in approach between the European style of psychiatry using a phenomenological approach, and North American psychiatrists who derived their ideas from psychoanalytic theory and thus used a much broader category.

This difference was clearly demonstrated by the US–UK diagnostic project, and later by the International Pilot Study on Schizophrenia which showed that of the nine centres in the study seven agreed with each other (Denmark, India, Columbia, Nigeria, England, Taiwan, Czechoslovakia) whilst the other two, the USSR and USA used a wider concept (World Health Organization, 1973).

5.5 Standardized diagnostic schedules

Comparison between the diagnoses made in the above studies was possible by using a standardized interview schedule, the Present State Examination (P.S.E.) which is a clinical interview with standardized criteria for rating 140 items (Wing, Cooper and Sartorius, 1974). The scores are then rated on a computer programme (CATEGO) to provide different classes and subclasses which correspond closely to accepted diagnostic categories. The CATEGO programme, like conventional diagnosis, places weight on the first rank symptoms. Although not described as such by its authors it operates in a way similar to the 'operational criteria' recently developed in a variety of centres, mostly North American. These operational criteria differ from the conventional classifications in that they depend not on a categorical system, which provides descriptions of typical illness, but no indication of how a less than typical case should be classified, but on a 'flow chart' method. This provides obvious advantages. To quote Kendal (1972):

> 'It is not enough to say "the typical features of schizophrenia are A, B, C and D." The statement must be recast in some form such as "For a diagnosis of schizophrenia at least two of A, B, C and D must be present and E must be absent."'

One of the earliest of these was provided by Feighner et al. (1972).

> 'A through C are required:
> A: Both of:
> > (i) A chronic illness with at least 6 months of symptoms prior to the index evaluation without return to the premorbid level of psychosocial adjustment.
> > (ii) Absence of a period of depressive or manic symptoms sufficient to qualify for a diagnosis of affective disorder or probable affective disorder (defined elsewhere).
> B: At least one of:
> > (i) Delusion or hallucination without significant perplexity or disorientation associated with them.
> > (ii) Verbal production making communication difficult because of a lack of logical or understandable organization.
> C: At least three of the following must be present for a diagnosis of 'definite' schizophrenia and two for a diagnosis of 'probable' schizophrenia.
> > (i) Single
> > (ii) Poor premorbid social adjustment or work history.
> > (iii) Family history of schizophrenia.
> > (iv) Absence of alcoholism or drug abuse within one year of onset of psychosis.
> > (v) Onset of illness prior to age of 40.'

Other diagnostic schedules have followed including the Research Diagnostic Criteria of Spitzer et al. (1975):

> 'A through C are required for the episode of illness being considered.
> A At least 2 of the following are required for definite, and 1 for probable:
> > (i) Thought broadcasting, insertion, or withdrawal (as defined in this manual).

   (ii)   Delusions of control, other bizarre delusions, or multiple delusions (as defined in this manual).

   (iii)   Delusions other than persecutory or jealousy, lasting for at least one week.

   (v)   Auditory hallucinations in which either a voice keeps up a running commentary on the patient's behaviours or thoughts as they occur, or 2 more voices converse with each other.

   (vi)   Nonaffective verbal hallucinations spoken to the subject (as defined in this manual).

   (vii)   Hallucinations of any type throughout the day for several days or intermittently for at least 1 month.

   (viii)   Definite instances of formal thought disorder (as defined in this manual).

   (ix)   Obvious catatonic motor behaviour (as defined in this manual).

B A period of illness lasting at least 2 weeks.

C At no time during the active period of illness being considered did the patient meet the criteria for either probable or definite manic or depressive syndrome (criteria A and B under Major Depressive or Manic Disorders) to such a degree that it was a prominent part of the illness.'

Although some of these symptoms are subjectively assessed they are well defined, and the form of arriving at a diagnosis is straightforward, allowing high reliability of diagnosis. As they define clear populations most psychopharmacological research makes use of them. However, from the above two examples it may be seen that there are clear differences between them: the length of illness being six months in the first case and two weeks in the second, and social and aetiological factors being taken into account in the first, whereas the second is based solely on clinical symptoms. The population studied will vary with the definition used.

5.6 Outcome

The acute symptoms of schizophrenia are not normally long lasting, and particularly since the introduction of the phenothiazine and related drugs (see Section 5.13 on) rapid response of symptoms to treatment is the rule. Some patients respond poorly or not at all to such medication but the symptoms may eventually become less prominent as the illness remits naturally.

Recurrent episodes or relapse of illness are common, sometimes apparently precipitated by failure to take medication, but often for some unknown reason. The introduction of long acting 'depot' preparations of neuroleptic drugs (Section 5.14) which are given by intramuscular injection, usually every 2–4 weeks, has proved successful in maintenance treatment, avoiding some of the problems of patient compliance.

The long term outcome is more variable and interpretation of this is somewhat confused. Some patients have an acute schizophrenic illness, recovering completely and remaining well. At the other extreme some patients never fully recover, have frequent relapses, and suffer socially from failure to keep or make personal relationships or from failure to maintain employment. All shades in between are seen and studies have attempted to identify prognostic features.

A good prognosis is associated with:

Absence of family history of schizophrenia
Stable premorbid personality
Warm personal relationships
Stable home relationships
Identifiable precipitatory factor
Rapid onset
'First-rank' symptoms not prominent
Intercurrent mood disturbance
Initiative, interest, and motivation retained
Appropriate treatment

It has been seen above that these features are taken into account to different extents by different diagnostic criteria, and so the outcome is to some extent dependent on the criteria used. Brockington, Kendell and Leff (1978) have demonstrated that the outcome in terms of incomplete recovery from the initial episode, persistence of symptoms, and social outcome, is predicted less well by some criteria than others, but it is not clear what are the features responsible for this.

5.7 Assessment of outcome

From what has been described above, it may be seen that the quantification, or assessment of outcome, of schizophrenia is complicated. On the one hand a measurement of mental state may be made and on the other an assessment of social functions.

Mental state. The most commonly used scale is the Brief Psychiatric Rating Scale (B.P.R.S.) developed by Overall and Gorham (1962) which rates 16 items on a scale from 0 to 7. These items are based on observation (tension, emotional withdrawal, mannerisms and posturing, motor retardation, uncooperativeness), or on verbal report (conceptual disorganization, unusual thought content, anxiety, guilt, grandiosity, depressive mood, hostility, somatic concern, hallucinatory behaviour, suspiciousness, and blunted affect). A 'total pathology' score is derived by adding the individual ratings made according to a well defined glossary. The symptoms are usually given equal weightings, as validation in terms of agreement with psychiatric assessments has shown unequal weightings to be of value only for specific subcategories of schizophrenia.

Readmission to hospital. This is a crude measure susceptible to many influences and of doubtful meaning.

Social functioning. This may be assessed by some major objective criteria, such as type of accommodation, having a job, readmission or rereferral rate, trouble with the authorities, or by some more detailed assessment of activity and interpersonal relationships.

5.8 Aetiological theories

It is our assumption that there is a biochemical basis for schizophrenia, and the direct evidence for this is reviewed later on in this chapter.

Such a biochemical change might be of genetic origin and considerable research has been directed towards the inheritable risk of developing schizophrenia. Whilst there is a 1% chance of an individual from the general population developing schizophrenia this increases to approximately 40% if both parents are affected, 10% if one parent or a sibling has the illness, and 2–3% for uncles, nieces, and first cousins. If one of a pair of twins has schizophrenia the chance of the other in dizygotic twins developing the illness is the same as in non-twin siblings. However, if one member of monozygotic twins is schizophrenic the expectation of the other developing the illness is 60%. These figures could of course be explained to some extent by environmental factors, but studies on adoptive children, and on monozygotic twins reared apart, both of which separate the effects of upbringing from heredity, support the assertion that there is a strong genetic component.

Recent studies have looked at the social factors referred to under outcome above, and have demonstrated an apparent association between life events and breakdown, or relapse. Higher relapse rates are also found in patients with high emotional involvement with their families, and with greater time spent in contact with them. Those patients with the greatest involvement seem to be the ones who benefit most from treatment. It is tempting to infer a causal relationship between these but the data may indicate that different types of schizoprenia generate different behaviours and show different responses to treatment.

PART 2   THE BIOCHEMISTRY OF SCHIZOPHRENIA

5.9   Endogenous psychotogens

Thirty years ago Osmond and Smythies (1952) proposed that patients with schizophrenia might be synthesizing endogenous hallucinogenic substances, probably by transferring methyl groups to the biogenic amines. Methoxylation of dopamine produces 3,4-dimethoxyphenylethylamine, which has been reported to produce catatonia in animals. Furthermore several methylated biogenic amines are hallucinogenic in man and administration of methionine (which forms *S*-adenosylmethionine, a methyl donor) makes some schizophrenic patients worse.

In 1962 there seemed to be a dramatic breakthrough. Friedhoff and Van Winkle (1962) found a spot on paper chromatograms of the urine of schizophrenics which was not found in the urine of the normal population – the 'Pink Spot'. It was identified as 3,4-dimethoxyphenylethylamine. However, not only was this compound not found to be active in humans but was also subsequently found to be excreted by normals, living in the same conditions as the schizophrenia patients, and was dietary in origin. The finding which was widely hailed at the time, remains an object lesson for the use of controlled studies.

Subsequently much effort was expended in looking for methylated indoles in urine, plasma, or CSF of schizophrenia patients. However, despite several optimistic initial reports nothing definite has emerged.

All of the psychotomimetic substances suggested, such as 5-methoxy-N,N-dimethyltryptamine and dimethyltryptamine, produce abnormal behaviours in animals by stimulating 5-HT receptors. It thus seems unnecessary to propose the existence of endogenous psychotogens as it is possible that schizophrenia could be caused by abnormal functioning of these normal 5-HT or catecholamine pathways – a far simpler proposal than that of abnormal psychotogen formation.

Hollister (1977) has argued against the concept of endogenous psychotogens, claiming that most experienced clinicians can distinguish between psychosis produced by hallucinogens and schizophrenia. As he says, the idea of endogenous psychotogens has been around for a long time but that in psychiatry old hypotheses never die, they simply appear in a new guise.

5.10 Monoamine neurotransmitter systems

Until recently little evidence has been available to indicate changes in the brains of schizophrenic patients.

The concentrations of 5-HIAA, HVA, and MOPEG in the CSF have been reported to be no different from controls. In the last 3–4 years, however, there have been some very comprehensive studies on the biochemistry of schizophrenia, mainly from post-mortem tissues which have indicated the existence of changes which may have importance in the pathology. Crow and colleagues conducted an extensive study of monoamine systems in post-mortem brains from schizophrenic patients (Crow *et al.*, 1979a) and it is worth reporting these findings in some detail.

Dopamine and its metabolites HVA and DOPAC were measured in n. caudatus, n. accumbens, and putamen. There were no changes in HVA or DOPAC but a modest increase in dopamine in the caudate, in contrast to an earlier finding of an increase in DA in the accumbens but not in the caudate (Bird *et al.*, 1977). The concentration of NA and 5-HT were raised in the putamen of schizophrenic patients. However, overall there is little evidence for any major abnormality of 5-HT systems (Joseph *et al.*, 1979).

The study also examined the activity of a range of monoamine-metabolizing enzymes. Dopamine-β-hydroxylase, tyrosine hydroxylase, and catechol-O-methyltransferase activities were unchanged in a variety of brain regions. The activity of MAO measured to four different substrates was very similar in 14 brain areas of control and schizophrenic brains.

Various points arise from this study. Whilst several studies have found no change in the activity of MAO in the platelets of patients with schizophrenia, there have been at least eight reports of lowered activity. The suggestion was made that this decrease might reflect a central abnormality. However the post-mortem data fail to support this suggestion.

Another point is the apparent absence of major monoamine changes in the brains of those subjects studied, despite the fact that they were on drugs before their death. However, recent animal data indicate that long term neuroleptic medication is unlikely to produce major changes in DA turnover. Of course it is difficult to assess DA turnover on the evidence of HVA concentrations since turnover is, by definition, a dynamic measurement.

Undoubtedly there are difficulties in interpreting post-mortem data: the complications of post-mortem changes, storage conditions, matched controls (age, sex, time of freezing the brain after death, etc.) and the effects of long term drug administration being obvious (see Chapter 1). Nevertheless the study reviewed here was thorough and carefully conducted and therefore provides formidable evidence against major metabolic changes in brain monoamine metabolism in schizophrenia.

However, these workers and others have provided other data suggestive of postsynaptic changes in dopamine function in schizophrenia. Ligand binding studies to assess the number and affinity of dopamine receptors in post-mortem schizophrenic brain tissue have been made. This technique has been discussed in Section 2.14.

The binding of spiroperidol (spiperone) was found to be greater in the brain of schizophrenic patients compared to the controls in those dopamine rich areas studied, namely the caudate nucleus, putamen, and nucleus accumbens. Scatchard analysis demonstrated that there was an increase in binding sites (Fig. 5.1). This finding has been confirmed by two other laboratories to date.

A major problem of interpretation is that neuroleptic administration itself increases the number of binding sites. Owen et al. (1978) therefore examined the medication of the schizophrenics up to death (Table 5.1). It can be seen that the increased binding may well not be the result of neuroleptic administration although there is clearly a greater increase in those subjects who had been treated with neuroleptics. One argument advanced by this group against the view that the changes are merely a drug induced effect is that the binding of the rigid dopamine agonist ADTN is not increased in post-mortem brains of schizophrenic patients (Owen et al., 1981). In contrast, chronic neuroleptic administration in animals also increases the binding of ADTN. One might still ask why neuroleptic administration to some of the subjects did not cause this change. However, a more major question arises; that is, if spiroperidol binding increases, but ADTN does not, what receptor type is increased in schizophrenia? It may well be that only one type of dopamine population is being changed (see Kebabian and Calne, 1979) or that these changes reflect a differential and unsuspected change.

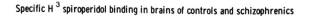

Specific H ^{3} spiroperidol binding in brains of controls and schizophrenics

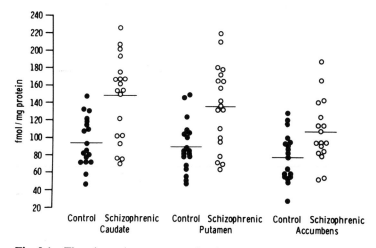

Fig. 5.1 The dopamine receptor density, assessed by spiroperidol binding in caudate, putamen, and n. accumbens in control and post-mortem schizophrenic brains. Differences between control and schizophrenic brains $P < 0.001$ for caudate and putamen and $P < 0.01$ for accumbens. (Reproduced from Owen *et al.* (1878) with permission of *The Lancet* and the authors)

Table 5.1 Relation between maximum specific [^{3}H]-spiroperidol binding and medication

Group	n	Binding (fmol/mg protein)	P
Controls	15	167 ± 50.4	—
Schizophrenics			
Never had medication	2	296.9; 214.8	< 0.05
No medication for > 1 year before death	5	249.7 ± 68.1	< 0.01
Uncertain if medicated up to death	6	380.9 ± 90.6	< 0.001
Definitely receiving medication	4	390.3 ± 163.4	< 0.001
All	15	339.7 ± 119.6	< 0.001

Taken from Owen *et al.* (1978) with permission of *The Lancet* and authors.

It should not be assumed that the post-mortem changes are indicative of some primary pathological change. The altered receptor population may result from a long term change in the function of the dopaminergic systems due to alterations in other disordered neurotransmitter systems. In the same way the neuroleptics may be blocking the dopamine systems and may improve symptoms, but not return to normal the primary pathological disorder.

The changes in dopamine receptor numbers are therefore of interest, but at present their significance remains to be shown.

There has been one report of reduced GAD (glutamic acid decarboxylase) activity in post-mortem of schizophrenic patients' brains, suggesting a possible change in GABA function. However, other studies failed to confirm this. The probable reason for the reported change was the high incidence of death from bronchopneumonia in the schizophrenic group. Death from bronchopneumonia results in a lowering of GAD activity (see Crow, Johnstone and Owen, 1979).

5.11 The possible involvement of a virus

Recently there has been a report on the presence of a virus-like agent in the CSF of patients with schizophrenia. One-third of the acute schizophrenic cases had the agent present in the CSF and it was also detected in patients suffering from a variety of neurological disorders. It was suggested that the virus has a wide distribution, but is pathogenic for schizophrenia only in a genetically disposed subpopulation (Tyrrell *et al.*, 1979; Crow *et al.*, 1979b). An extension of these data is clearly needed.

PART 3   THE PHARMACOLOGY OF SCHIZOPHRENIA

5.12   Introduction

Although the historical physical treatments of schizophrenia may have had pharmacological effects, the use of drugs was not conceptualized as producing a biochemical change, but as a means of producing gross physical change, as in the case of inducing convulsions with metrazol and inducing coma with insulin.

The first modern drug to be used in a specific way was the 'parent' of several of the neuroleptic drugs, chlorpromazine. Phenothiazine had been introduced as an antihelminthinic and other derivatives were being investigated for use against protozoal infection. One such derivative was promethazine which had marked antihistaminic actions. Chlorpromazine was found in addition to lower blood pressure and temperature. It also potentiated anaesthetic action and was therefore incorporated with promethazine and pethidine in a 'cocktail' as an adjunct to anaesthesia. It was noticed that when the drug was given alone it produced a lack of interest although the patient remained alert. This observation led to the first use of chlorpromazine in schizophrenic patients (Delay, Deniker and Hart, 1952).

It was in the same year as this report that the active alkaloid of *Rauwolfia serpentina* was first isolated – reserpine. As stated in Section 3.11 the root of this plant had been used for several centuries, and was claimed to be useful for both insanity and high blood pressure. In 1954 Weber compared reserpine with chlorpromazine, and referred to these two drugs as neuroleptics. Neuroleptic translated literally means acting on nerves, but it was and is used to mean drugs with therapeutic efficacy in the major psychotic or overactive psychiatric illnesses. Major tranquillizers is another term sometimes used for these drugs, to distinguish them from the minor tranquillizers, such as the benzodiazepines.

5.13 The neuroleptic drugs

The term covers a wide range of compounds, and the number available has increased steadily over the years.

The major group are the phenothiazines (Fig. 5.2), of which there are three main subgroups, classified by the side chain attached to the 10 position (marked X in Fig. 5.2).

1. Aminoalkyl compounds, for example chlorpromazine, which are relatively sedative and produce moderate extrapyramidal side effects.
2. Piperazine compounds, for example trifluoperazine, which are more potent and less sedative than group 1 but have a greater tendency to produce extrapyramidal side effects.
3. Piperidine compounds, for example thioridazine, which are approximately equipotent with group 1 but produce fewer extrapyramidal side effects.

The thioxanthine compounds are closely related to the phenothiazines, differing only in having a carbon instead of nitrogen atom at the 10 position (Fig. 5.3). This also confers a double carbon bond on the molecule and thus in the case of flupenthixol stereoisomeric forms exist (see Sections 5.16 and 5.17). Their properties are very similar to the respective phenothiazines.

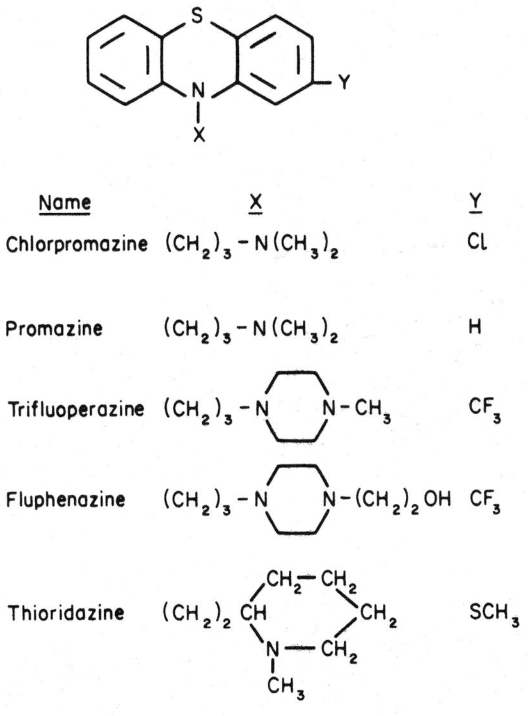

Name	X	Y
Chlorpromazine	$(CH_2)_3 - N(CH_3)_2$	CL
Promazine	$(CH_2)_3 - N(CH_3)_2$	H
Trifluoperazine	$(CH_2)_3 - N\bigcirc N - CH_3$	CF_3
Fluphenazine	$(CH_2)_3 - N\bigcirc N - (CH_2)_2 OH$	CF_3
Thioridazine	$(CH_2)_2\ CH\big\langle{}^{CH_2-CH_2}_{N-CH_2}\big\rangle CH_2$, $N-CH_3$	SCH_3

Fig. 5.2 The structures of five phenothiazines, showing examples of the three main groups

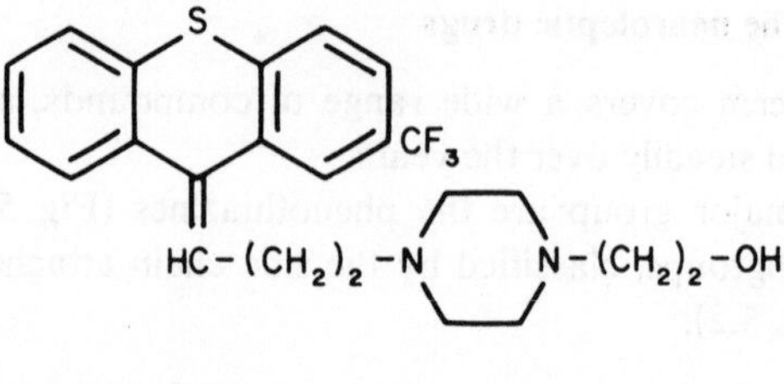

cis – Flupenthixol

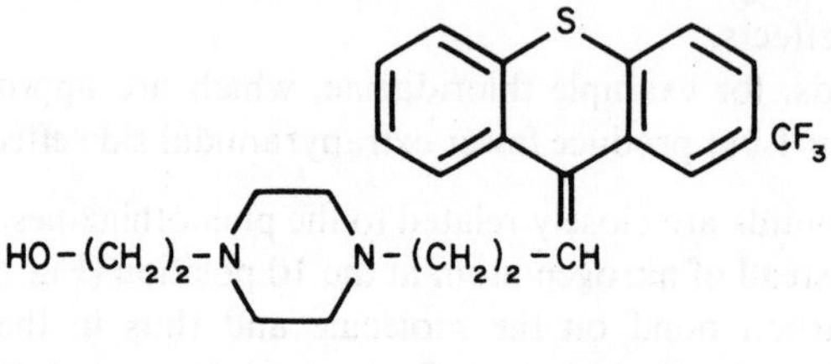

trans – Flupenthixol

Fig. 5.3 Structures of *cis*-(or α)-flupenthixol and trans-(or β)-flupenthixol

Other widely used compounds are the butyrophenones (Fig. 5.4), for example haloperidol. These are more potent than chlorpromazine but produce more extrapyramidal side effects. Derived from this group are the diphenyl-butylpiperidines, e.g. pimozide, which have a longer half-life than the other groups and need to be given only once daily.

Clozapine, a dibenzazepine (Fig. 5.4), was introduced as a therapeutically effective drug free from actions on the nigrostriatal system and thus free from Parkinsonian side effects. Whilst other side effects have necessitated its withdrawal from clinical use, it has been a useful drug in elucidating the therapeutic site of action of the neuroleptics (Section 5.18). It is sometimes referred to as an 'atypical neuroleptic' because of the absence of Parkinsonian side effects.

5.14 Depot neuroleptics

A problem with some schizophrenic patients is that they are unreliable in taking their medication. Depot preparations are now available to eliminate this problem and also to deal with patients who have poor absorption of oral drugs.

The drug in the form of a salt is dissolved in a viscous liquid, such as sesame oil, which is injected intramuscularly. The drug is then slowly released from the intramuscular site and metabolized to the free drug. Treatment time varies with the preparation but is usually once every 1–4 weeks and very steady plasma levels of the drug are seen throughout this time.

Butyrophenones

Spiroperiol

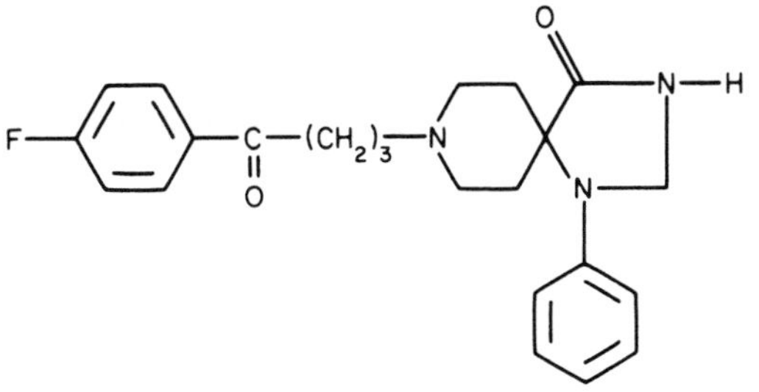

Haloperidol

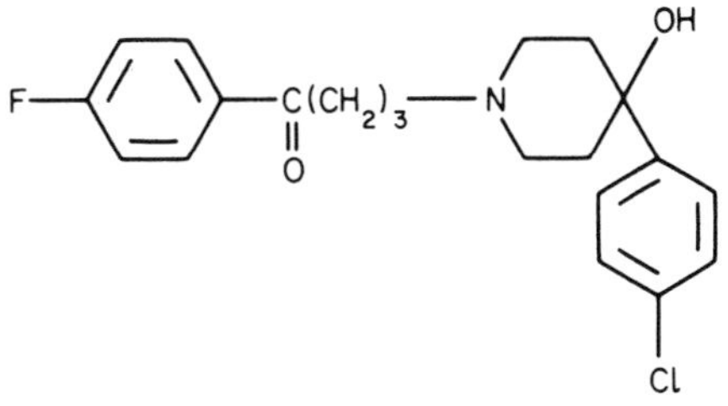

Pimozide

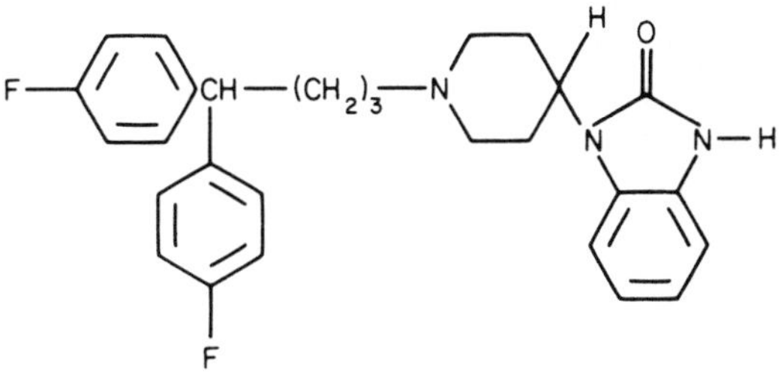

Dibenzazapine

Clozapine

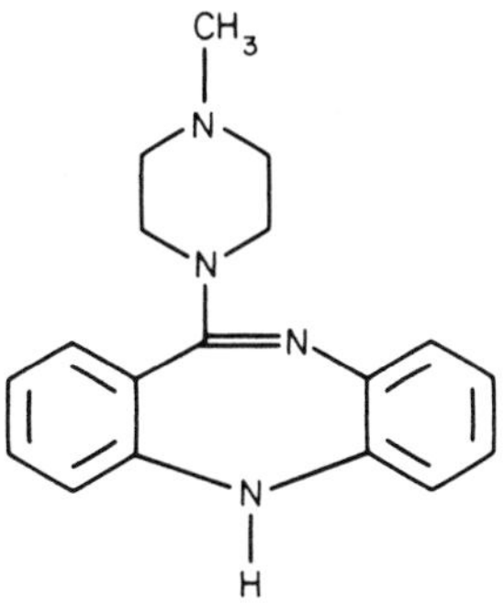

Fig. 5.4 The structures of three common butyrophenones and a dibenzapine, clozapine

Drugs available for use this way include fluphenazine (as decanoate or enanthate salts) and flupenthixol (as decanoate).

5.15 Action of neuroleptics on dopamine metabolism and behaviour

Clearly it would strengthen any theory concerning the aetiology or treatment of schizophrenia if all these apparently diverse drugs were shown to have some similarity in mode of action. The first clue was provided in 1963 in Göteborg when Carlsson and Lindqvist demonstrated that whilst chlorpromazine and haloperidol did not alter rat brain dopamine concentrations, they did increase the concentration of the major dopamine metabolite HVA whilst leaving noradrenaline metabolite concentrations unchanged. It has subsequently been found that essentially all neuroleptics, regardless of structure, produce this effect. The combination of normal amine and raised metabolite concentration suggested increased turnover of dopamine, and using various techniques this interpretation has been confirmed.

Why should the drugs increase turnover? An indication was provided by the action of these drugs on the behavioural changes induced in animals by amphetamine.

Amphetamine is structurally related to the catecholamines (see Fig. 5.5) but can enter the brain. Administration to rats of amphetamines at low doses produces locomotion and at higher doses various stereotyped behaviours including licking, biting, sniffing, and chewing. Such behaviours have been shown to be due predominantly to release of dopamine from nerve endings, rather than 5-HT, which it can also release. In some elegant studies it has been demonstrated that the locomotion is primarily due to release of dopamine in the mesolimbic forebrain (nucleus accumbens or A10 system) whilst the stereotypy is the result of dopamine release in the caudate nucleus (or A9 system) (Kelly, Seviour and Iversen, 1975).

This amphetamine behavioural syndrome in animals is inhibited by neuroleptic drugs which also indicates that these drugs interact in some way with dopamine systems. Together the data suggest that neuroleptic drugs are 'blocking' the dopamine receptor. Such an action would both inhibit the amphetamine behaviour

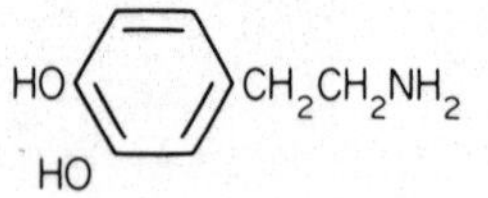

Dopamine

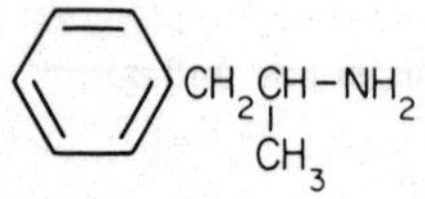

Amphetamine

Fig. 5.5 Similarity in structure between dopamine and amphetamine

and result in increased DA turnover. This latter action would result from a blockade of the presynaptic dopamine receptors which are involved in regulating turnover (see Section 2.11).

Other observations with amphetamine produced a further link with schizophrenia. In 1958 Connell stated that the syndrome induced by large doses of amphetamine closely resembled acute paranoid schizophrenia, an observation that has been supported by various other groups. It has been shown that this 'amphetamine psychosis', like the amphetamine induced behaviour in rats is reversed by chlorpromazine.

Several other clinical findings also point to the view that neuroleptics decrease dopamine activity. Prolactin release from the pituitary is inhibited by the dopamine containing tuberoinfundibular system. It is now thought that prolactin release inhibiting factor is dopamine itself, or that dopamine is the final control on this inhibitory factor. Neuroleptic drug administration results in a raised plasma prolactin concentration, as would be expected if neuroleptics block dopamine receptors in the pituitary.

Administration of most neuroleptics can give rise to extrapyramidal or Parkinsonian-like side effects and it is now established that in Parkinson's disease there is a marked decrease in the concentration, and therefore presumably the function, of dopamine in the substantia nigra (see Section 8.7).

Finally, as stated earlier, reserpine has been used with success in treating schizophrenia. It is known that this drug decreases brain monoamine concentrations, including dopamine.

5.16 Dopamine-sensitive adenylate cyclase

The next major advance in the understanding of the action of the neuroleptics came in the early 1970s predominantly from the laboratory of Greengard at Yale and subsequently from that of Iversen in Cambridge.

In homogenates of brain striatal tissue it had been demonstrated that addition of dopamine increased the formation of cyclic AMP, the so-called 'second messenger' (see Section 2.13). Greengard's and Iversen's groups showed that the neuroleptics inhibited the formation of cyclic AMP in brain homogenates, that is, these drugs inhibited dopamine-sensitive adenylate cyclase. The inhibition is dose dependent and the inhibition by the drug is competitive with dopamine.

In general, there is a degree of correlation between the therapeutic action of the drugs and their ability to inhibit dopamine-sensitive adenylate cyclase. Thus those related drugs which lack antipsychotic activity but are used as antiemetic or antihistaminic drugs (the phenothiazines promethazine or ethopropazine for example) or anti-anxiety or antidepressant drugs do not inhibit adenylate cyclase, or do so very weakly. A further observation was that structural analogues of some drugs differed markedly in their potency. For example α- or cis-flupenthixol was a potent adenylate cyclase inhibitor whilst the β- or trans-form was not (see Iversen, 1975a). Antischizophrenic activity has been shown to be due to the α-form (Section 5.17).

Much of the data discussed above have been reviewed by Iversen (1975a) where references to the original work will be found.

However, some discrepancies became apparent. The butyrophenones are extremely potent antipsychotic drugs, and are also potent drugs in terms of blocking dopamine-mediated behaviours in rats (such as amphetamine or apomorphine induced locomotor activity). However in terms of their ability to inhibit dopamine-sensitive adenylate cyclase their activity is a little less than chlorpromazine, a much less potent drug both clinically and experimentally. Of course it is possible that this lack of correlation between *in vitro* and experimental and clinical efficacy was in some way related to a pharmacokinetic phenomenon of absorption, distribution etc., but examination of the correlation between effects of neuroleptics on adenylate cyclase and experimental and clinical efficacy revealed that the correlation is weak when a large group of drugs is assessed (Fig. 5.6).

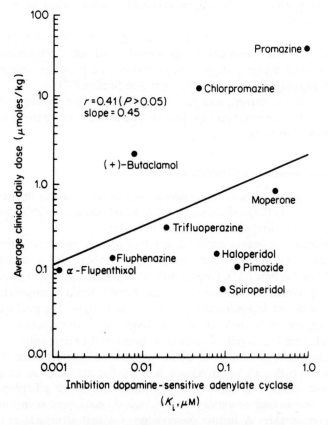

Fig. 5.6 Correlation between the affinities of anti-schizophrenic drugs for inhibition of dopamine-sensitive adenylate cyclase and antagonism of apomorphine stereotyped behaviour in the rat. (Reproduced from Snyder *et al.* (1977) with permission of the Oxford University Press)

5.17 Ligand binding studies

These discrepancies between the *in vitro* and *in vivo* data on neuroleptics which weakened the view that the mechanism of the drugs action was simply by inhibiting adenylate cyclase were further investigated by use of an exciting new neuropharmacological approach which was being developed and exploited by Snyder's department at Johns Hopkins in Baltimore. The techniques was that of ligand binding which allows as assessment of receptor number and function to be made (see Section 2.14).

The techniques have been used to examine the affinity of neuroleptics for the dopamine receptor. Data obtained have demonstrated a good relationship between the dose required to inhibit dopamine-mediated behaviour in rats and the dissociation constant (K_D; which is 1/affinity constant) of the drugs at the

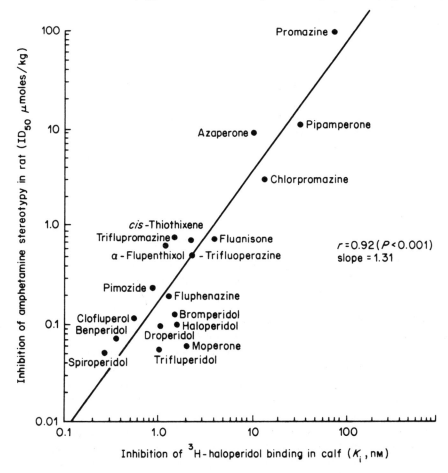

Fig. 5.7 Correlation between affinities of antischizophrenic drugs for [³H]-haloperidol binding site and antagonism of amphetamine stereotypy in the rat. (Reproduced from Snyder *et al.* (1977) with permission of the Oxford University Press)

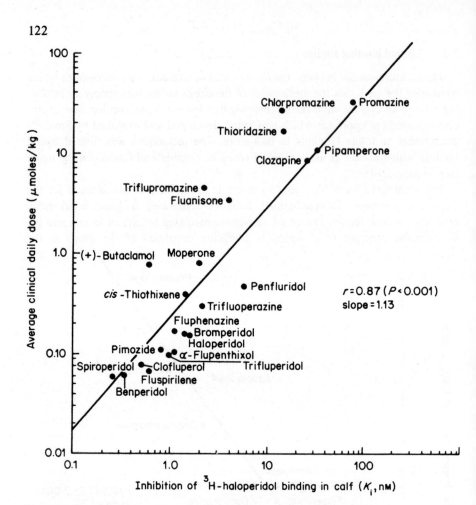

Fig. 5.8 The correlation between the affinities of antischizophrenic drugs for [³H]-haloperidol binding site and clinical potencies. (Reproduced from Snyder *et al.* (1977) with permission of the Oxford University Press)

dopamine receptor (Fig. 5.7). Similarly a reasonable relationship has been found between clinical potency and dissociation constant (Fig. 5.8).

Clearly there are problems in experiments using ligand–receptor binding and these have been outlined in Section 2.14. It should be pointed out that to plot 'average daily clinical dose' ignores the large variations in doses given which are determined by the nature of the illness and the variation in rates of metabolism of some of these drugs.

Furthermore some of these drugs (chlorpromazine is the prime example) have a variety of metabolites, some of which are almost certainly therapeutically active. Nevertheless a plot such as that shown in Fig. 5.8 does indicate some relationship.

At present therefore the general view is that the clinical action of neuroleptics may acutely inhibit dopamine neurotransmission. Whether this is how they act in the long term is open to some doubt and is discussed later (Section 5.21).

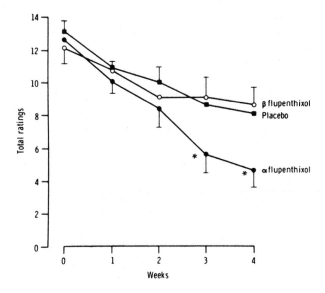

Fig. 5.9 Improvement in total symptoms as a function of treatment with α- and β-isomers of flupenthixol and placebo *P < 0.05 vs p-flupenthixol or placebo. (Reproduced from Johnstone *et al.* (1978) with permission of *The Lancet* and the authors)

That dopamine antagonism is important for the action of the drugs has recently received support from a clinical investigation. As stated earlier the drug flupenthixol exists in two isomeric forms (Fig. 5.3). Animal data have indicated clearly that the *cis*- or α-form blocks dopamine-sensitive adenylate cyclase and dopamine receptors while the β- or *trans*-form is essentially without effect. The preparation of flupenthixol used clinically is a 50–50 mixture of the isomers. Johnstone *et al.* (1978) undertook a therapeutic trial of the two isomers of flupenthixol in a group of acute schizophrenics. In addition to the two drug groups a placebo group was included. During the course of the trial chlorpromazine was given when necessary. Some of the results of the trial are shown in Fig. 5.9. The quite marked improvement in the placebo group should be noted – probably due in part to the care and attention being given to the subjects. It is clear that the β- or inactive isomer is no better than placebo and that after 2 weeks improvement has occurred only in those patients given the α-isomer. In addition the β-isomer and placebo treated patients needed significant quantities of chlorpromazine during the course of the investigation. It is possible that the α-form is more potent than the β-form in some other pharmacological property, but there seems to be little indication for this (see Johnstone *et al.*, 1978) and we must therefore conclude that the clinical potency of α-flupenthixol is probably a consequence of its dopamine antagonist actions.

One further point to be made is that the α-flupenthixol was most effective in improving the 'positive' schizophrenic symptoms (hallucinations, incoherence, delusions) and had little effect on either 'negative' symptoms (flattening of affect and poverty of speech) or the non-specific symptoms such as depression, anxiety, or retardation.

5.18 The mesolimbic forebrain

Where is the neuroanatomical site of action of the drugs? Most of the neuroleptics produce extrapyramidal side effects, which are probably due to dopamine antagonist actions in the corpus striatum. However, extrapyramidal side effects do not correlate with clinical efficacy; for example thioridazine is equipotent with chlorpromazine in clinical potency, but produces considerably fewer Parkinsonian side effects. One explanation advanced was that some of the neuroleptics had greater anticholinergic potency than others and Miller and Hiley (1974) were able to show an inverse correlation between the anticholinergic potency of various drugs and their ability to produce extrapyramidal side effects. Thioridazine has potent anticholinergic activity, providing a plausible explanation for the absence of Parkinsonian side effects. It also poses a problem, however, since one might therefore suggest that this action would prevent the therapeutic action by antagonizing the antidopaminergic effect. Furthermore Crow, Johnstone and McClelland (1976) have reported cases of patients with both Parkinson's disease and schizophrenia. If schizophrenia were simply the result of 'overactivity' of dopamine systems in the striatum, one would expect Parkinsonism and schizophrenia to be mutually exclusive.

Reference to Section 2.7 shows the existence of three distinct dopamine systems in the brain: the nigrostriatal, the mesolimbic forebrain and the tuberoinfundibular systems. The last named system is involved in control of prolactin secretion, exerting an inhibitory effect on release of this hormone. Since studies have not suggested a decreased release of prolactin in schizophrenics, there does not appear to be an abnormality of this dopamine system.

The mesolimbic forebrain dopamine system (the A10 system) arises from cell bodies near the substantia nigra and projects to the nucleus accumbens. This system has connections with the frontal cortex, olfactory tubercle, hippocampus, and caudate nucleus.

Crow, Deakin and Longden (1975) investigated the hypothesis that neuroleptic drugs produce their therapeutic effect by acting on the mesolimbic forebrain, by examining the effect of three drugs on dopamine turnover, as assessed by the rise in HVA in both the n. accumbens and striatum. In Fig. 5.10 it can be seen that when given to rats at doses equivalent to those given clinically the drugs fluphenazine, chlorpromazine, and thioridazine had markedly different effects on turnover in the striatum. There was a good correlation between the alteration of turnover and the potency of the drugs in producing extrapyramidal side effects, fluphenazine having a high incidence and thioridazine low. However in the n. accumbens the drugs display equal effects, perhaps reflecting the equivalence between clinical potencies at the doses chosen.

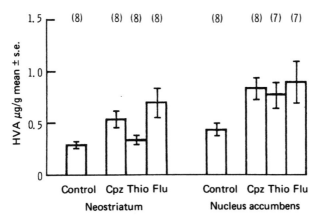

Fig. 5.10 Effects of three phenothiazines given at different doses equivalent to their therapeutic effects on the HVA concentration in the striatum and nucleus accumbens. Doses: chlorpromazine (Cpz 5 mg/kg), thioridazine (Thio 5 mg/kg) and fluphenazine (Flu 0.25 mg/kg). (Reproduced from Crow, Deakin and Longden (1975) with permission of Macmillan Journals Ltd)

Other data also indicate that the antipsychotic action of the neuroleptics is the result of dopamine antagonist actions in the accumbens. Amphetamine reduces the firing rate of both striatal and mesolimbic dopamine systems. Thioridazine (in line with its lack of effect in altering DA turnover in the striatum) and clozapine have little effect in preventing this reduction in firing. However, both drugs prevent the shut down in firing in the mesolimbic forebrain, which would fit with the fact that both drugs are antipsychotic but produce few extrapyramidal side effects.

In experimental animals it has been suggested that amphetamine induced locomotor activity is the result of DA release in the accumbens whilst the behavioural stereotypy seen after higher dose is striatally mediated (Kelly, Seviour and Iversen, 1975). Clozapine has little effect on stereotypy.

Other studies on thioridazine, clozapine, and sulpiride using techniques of injection at specific sites in the brain have indicated the relative potency of these compounds to inhibit dopamine-mediated behaviour initiated by the n. accumbens and the failure of the non-antipsychotic drug metoclopramide to block this behaviour (Costall and Naylor, 1977).

5.19 Tardive dyskinesia

The foregoing data therefore suggest that in experimental animals neuroleptics are dopamine antagonists and that their clinical efficacy on initial administration is also probably the result of dopamine antagonist actions. Furthermore, the probable site of action for their antipsychotic effect is the mesolimbic forebrain whilst dopamine antagonism in the caudate nucleus is responsible for production of extrapyramidal side effects. The problem is that most of these studies have been either *in vitro* investigations or short term studies in either animals, or in the case of the flupenthixol trial (Johnstone *et al.*, 1978), human subjects. Whether dopamine antagonism is the way the drugs act during longer term administration is a different question. Even the 'acute' flupenthixol trial raises problems insofar as dopamine blockade – as assessed by the rise in plasma prolactin – occurred very rapidly whilst the antipsychotic effect took about three weeks.

When people are treated over several years with neuroleptic drugs a proportion develop tardive dyskinesia. The dyskinesias resemble those produced by L-dopa administration (Section 8.13), can be reduced by increasing the neuroleptic dose, and are often exacerbated by neuroleptic withdrawal (Marsden, Tarsy and Baldessarini, 1975). On the basis of such evidence it has been suggested that the dyskinesias result from dopamine overactivity possibly due to dopamine receptor supersensitivity resulting from the long term receptor blockade. Such supersensitivity changes can be shown in animals after 2–3 weeks' neuroleptic administration, but only after drug withdrawal. In contrast tardive dyskinesias are seen after months, or years of drug administration and whilst the patients are still taking the drugs.

Since the dopamine supersensitivity hypothesis of tardive dyskinesia is not directly compatible with the suggested dopamine antagonist role of these drugs in schizophrenia, this raises the question as to whether continued dopamine blockade occurs during long term neuroleptic therapy.

Recently there has been a major investigation in animals in an attempt to clarify the long term actions of the neuroleptics.

5.20   The effect of long term neuroleptic administration to rats

Rats were administered either trifluperazine or thioridazine in drinking water for 12 months. In summary, it was found that whilst stereotyped behaviour induced by apomorphine was initially inhibited, longer term administration (greater than 6 months) resulted in the inhibition being replaced by enhanced response (Fig. 5.11). Such changes occurred whilst the animals were still receiving the drug continuously. Biochemical data, measurement of dopamine-sensitive adenylate cyclase and dopamine–receptor ligand binding also showed that initial antagonist action of the drugs were, in time, replaced by enhanced responsiveness of the receptor systems (Clow, Jenner and Marsden, 1979).

Finally it should be noted that in these studies it was observed that the animals displayed spontaneous mouth movements after 12 months on the drugs. These, however, disappeared fairly soon after drug withdrawal, in contrast to clinical tardive dyskinesias, and so the relevance of these data to this drug induced side effect is uncertain. Nevertheless, the data do confirm the suggestion made by Klawans (1973) that in tardive dyskinesia receptor supersensitivity occurs during long term neuroleptic administration and whilst the drugs are still being received.

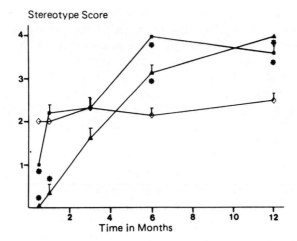

Fig. 5.11 The enhanced dopamine-mediated behaviour (apomorphine induced stereotyped behaviour) following long term neuroleptic administration ◇, control rats (distilled water); ■, thioridazine (30–40 mg/kg per day); ▲, trifluoperazine (2.5–3.5 mg/kg per day). *Different from controls $P < 0.05$. (Reproduced from Clow, Jenner and Marsden (1979) with permission of Elsevier-North Holland Ltd)

5.21 Dopamine antagonism as an explanation for the antipsychotic action of neuroleptics

The data of Clow and her colleagues raise the question as to whether neuroleptics act chronically by dopamine antagonism. Long term, parallel studies by the same group on the effects of the drugs on other neurotransmitters have so far not indicated major changes in other systems although Dawbarn, Long and Pycock (1981) have also shown 5-HT receptor supersensitivity. There are, however, some indications both that neuroleptics may influence the function of other neurotransmitters and that drugs thought not to act on dopamine neurotransmission are of value in schizophrenia.

The function of several other neurotransmitters has been shown to be altered during neuroleptic administration, though again it should be emphasized that these have been short term studies. Furthermore, it is probable that most of the changes seen are indirectly due to dopamine antagonism and result from the complex neurotransmitter interactions now being slowly elucidated.

The rate of acetylcholine turnover is decreased by haloperidol and chlorpromazine in several brain regions. Haloperidol and clozapine were also shown in the same study to decrease the turnover time of GABA in substantia nigra, caudate and n. accumbens (Marco *et al.*, 1976).

After eight weeks' treatment with haloperidol or chlorpromazine there is an increase in the number of GABA receptors in the substantia nigra but not striatum, indicating possible changes in the striatonigral GABA 'feedback' loop thought to regulate, in part, the nigrostriatal dopamine system. Of interest was the observation that clozapine did not have this effect (Gale, 1980).

Studies on met-enkephalin have indicated that repeated neuroleptic administration (over three weeks) increased turnover of this peptide (Hong *et al.*, 1978).

Hornykiewicz (1977) has proposed a possible role of noradrenaline in schizophrenia. Propranolol has been reported to be of value in treating schizophrenics (see Yorkston *et al.*, 1977) but the efficacy of propranolol as an antischizophrenic drug remains controversial and it is clear that even if its value is confirmed it will not replace conventional neuroleptic therapy. On the other hand it is interesting that propranolol is essentially devoid of antidopaminergic activity and so any therapeutic action is not due to dopamine antagonism. In the clinical trials conducted on this drug, large doses have been administered, and in animal behavioural studies propranolol in large doses can be shown to be a 5-HT antagonist (Green and Grahame-Smith, 1976). There is some evidence that even at quite low concentrations the drug is a 5-HT antagonist (Middlemiss, Blakeborough and Leather, 1977; Weinstock, 1980).

The fact that all clinically effective neuroleptics are dopamine antagonists, whilst structurally closely related compounds which are not dopamine antagonists are ineffective in the treatment of schizophrenia, suggests that *acutely* the therapeutic action of the neuroleptic drugs is by dopamine antagonism. This view is enhanced by the trial of the flupenthixol isomers (Section 5.17). Perhaps the simplest explanation for the action of these drugs when given chronically is that whilst some populations of dopamine receptors show supersensitivity after long-term neuroleptic treatment, others are enduringly antagonized. Certainly those involved in inhibition of prolactin release remain antagonized. Plasma prolactin levels remain elevated in most patients on chronic neuroleptic therapy. We may therefore hypothesize that those dopaminergic receptors involved in the anti-schizophrenic action of neuroleptics are also antagonized in the long term. Recent behavioural data with rats appears to support this contention. Gamble and Waddington (1981) have shown that after six months phenothiazine treatment to rats, some apomorphine induced behaviours are enhanced whilst others remain inhibited. Clearly, if this hypothesis is correct, it may be possible to develop drugs with anti-schizophrenic activity that do not produce tardive dyskinesias, on the assumption that the latter results from supersensitivity of a distinct class of dopamine receptors.

Nevertheless the data on propranolol taken with other data which fail to demonstrate a blockade of central dopamine systems during long-term neuroleptic administration mean that we should remain somewhat sceptical of the view that dopamine antagonist actions are either a necessary or sufficient requirement for anti-schizophrenic action.

Chapter 6

Alzheimer's Disease and Senile Dementia

PART 1 CLINICAL ASPECTS

6.1 Introduction

Dementia in the elderly is currently a major health problem in developed countries, and this is likely to increase in the future as a greater proportion of the population survives into old age.

There are thought to be several types of dementia, most of them clearly caused by specific factors such as infections, cerebral tumours, and multiple vascular infarct or arteriosclerotic dementia. Huntington's chorea is discussed in Chapter 7. In this chapter we are concerned with degenerative, or senile dementia, that is occurring in persons aged over 65, and Alzheimer's disease, a dementia in younger patients.

Bowen (1980) has pointed out that on neuropathological grounds there are no good reasons for separating presenile dementia or Alzheimer's disease from senile dementia (or senile dementia of the Alzheimer's type). The division is merely that of age, the former occurring before the age of 65, the latter in persons over that age. Apart from manifesting earlier, the presenile form, in general, has a more rapid progression.

We will use the term Alzheimer to denote both types, but in practice most studies have looked at the older group as that is more common.

6.2 Prevalence

It has been found that approximately 5% of people over 65 are demented. The incidence in those under 65 is very much less, perhaps 0.1% which represents some 50,000 people in the U.K.

6.3 Clinical features

The condition is characterized by a gradual loss of intellect and memory, and changes in personality.

129

Intellect. In the early stages this may be evident only as loss of interest and performance, but progresses to more obvious changes with loss of ability to understand or conceptualize problems with consequent loss of judgement.

Memory. Initial problems may be little more than a failure to remember names, or unimportant details, but progress to inability to remember recent events or current affairs, and later to loss of earlier memories, and disorientation. This leads to perseveration, or needless repetition of speech or acts, and confabulation where gaps in memory are wrongly filled in without the patient realizing the error.

Personality. Initially there may be only a coarsening of earlier personality traits. Later emotional disturbances with irritability, impulsiveness, anxiety, and depression are features with the patient moving from one to the other quickly and inappropriately. Suspiciousness may result from a sense that the patient is not in control of what is going on. In the later stages emotional blunting is more common.

As the above changes are the consequence of diffuse pathology, there is no consistent pattern to the development of the syndrome and a wide variety of clinical presentations is seen.

In the most severely afflicted patients specific or focal signs may be present. These include dyspraxia, difficulty performing fine movements such as dressing, dysphasia, an inability to express oneself verbally, and agnosia, an inability to recognize objects.

Although there are grounds for regarding the above syndrome as a form of accelerated ageing there are structural and biochemical reasons for regarding dementia as a distinct entity. In addition dementia appears to be more common in first degree relatives of affected persons, suggesting a genetic or possibly environmental factor.

PART 2   BIOCHEMISTRY

6.4   Pathological changes

Pathological changes occur in Alzheimer's disease and complicate biochemical studies. As in Huntington's chorea, degeneration complicates the interpretation of biochemical data which are normally reported in terms of weight of compound per unit weight of brain tissue. There are gross macroscopic changes in the brains of Alzheimer's patients; the weight of the brain can be decreased by up to one-third; the cerebral hemispheres show atrophic changes particularly the frontal and temporal lobes; and the lateral ventricles are enlarged and the white matter is reduced in volume (Tomlinson, 1980).

Microscopically there are characteristic changes – plaque formation and neurofibrillary tangles occurring throughout the cerebral cortex. Plaques are seen in silver impregnated histological slides prepared for light microscopy. They are irregular masses varying from 15–200 μm in diameter which may occupy much of the cortex, but can also be seen in hippocampus and amygdala. They consist of swollen nerve terminals, filled with laminated or dense, round or oval bodies which often contain abnormal fibrillary material (see Tomlinson, 1980).

Neurofibrillary tangles can also be seen under light microscopy using silver impregnation and appear as thick intracytoplasmic fibrils, curving around the nucleus and involved in the neuronal perikaryon.

There appears to be a loss of neuron cell bodies in the neocortex, although this is not as great as sometimes claimed when proper comparison is made with age matched controls.

Other pathological changes occur and readers interested should consult the recent review of Tomlinson (1980).

Dayan (1974) has pointed out that this widespread degeneration makes it unlikely that any single neurotransmitter is selectively affected, a view that can be supported by various studies.

6.5 Brain catecholamines

It is well established that both dopa decarboxylase and tyrosine hydroxylase activity decrease with old age, resulting in a decrease in catecholamine synthesis. Brain catecholamine concentrations also tend to decrease with age and therefore appropriate age matched controls are therefore vital in any biochemical studies.

In 1968 Gottfries, Gottfries and Roos showed that the dopamine metabolite homovanillic acid (HVA) was reduced in the basal ganglia of brains from

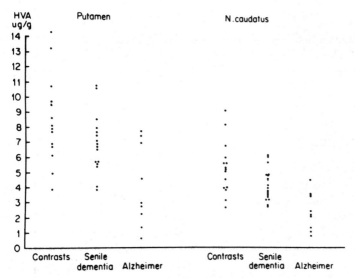

Fig. 6.1 HVA (μg/g) in caudate nucleus and putamen of three groups. Controls (contrasts), senile dementia, and Alzheimer's. (Reproduced from Gottfries, Gottfries and Roos (1968) with permission of Excerpta Medica)

Table 6.1 Levels of monoamines and metabolites (means $\pm$ s.e.m. nmol/g) in the caudate nucleus of cases with senile dementia and of control cases with no known neuropsychiatric disorder (data from an on-going Swedish study)

	Controls ($n = 17$)	Senile dementia ($n = 15$)
Dopamine	15.8[a]	8.5***
	$\pm$ 0.90	$\pm$ 1.62
3-Methoxytyramine	6.6[a]	4.7**
	$\pm$ 0.34	$\pm$ 0.99
Homovanillic acid	22.7	15.1**
	$\pm$ 1.37	$\pm$ 1.59
Noradrenaline	0.16	0.12*
	$\pm$ 0.012	$\pm$ 0.014
MOPEG	0.10	0.18*
	$\pm$ 0.018	$\pm$ 0.028
5-HT	0.51	0.25***
	$\pm$ 0.043	$\pm$ 0.040
5-HIAA	1.88	1.90
	$\pm$ 0.15	$\pm$ 0.028
Age (years)	72.8	75.7
	$\pm$ 1.90	$\pm$ 1.93

[a]$n = 54$; *$P < 0.05$; **$P < 0.01$; ***$P < 0.001$.
Reproduced from Gottfries (1980) with permission of John Wiley & Sons Ltd.

Alzheimer patients measured at post-mortem. This work was subsequently extended and lowered HVA concentrations were also shown in the caudate nucleus, putamen, and globus pallidus. Furthermore analysis of these data with the degree of dementia seen in the patients prior to death suggested that the greater the intellectual impairment the lower the HVA content. Subjects with pre-senile dementia had lower HVA than those with senile dementia (Fig. 6.1).

Subsequently the same group has also found lowered dopamine and noradrenaline levels in post-mortem brains from patients with presenile dementia compared with age matched controls.

Gottfries (1980) has recently quoted data from a current Swedish study on monoamine concentration in senile dementia cases which confirm the earlier studies (Table 6.1).

Measurement of CSF amine metabolites in Alzheimer type dementia patients have also been performed. Not surprisingly HVA concentrations are decreased and it was shown that dopamine synthesis is reduced using the probenecid test to measure synthesis rate.

Finally there is a report of lowered spiroperidol binding in caudate nucleus (Reisine *et al.*, 1978), suggesting a decrease in the dopaminergic receptor population.

6.6 Brain 5-hydroxytryptamine

In contrast to the changes in catecholamines it has been found that brain 5-HT content increases with age, at least in the mesencephalon and medulla oblongata. In the globus pallidus and hippocampus 5-HT does decrease with age. The major metabolite 5-HIAA shows a positive correlation with age in the caudate nucleus, hippocampus, and cerebellum (Gottfries, Gottfries and Roos, 1979).

In the Swedish study (Table 6.1) 5-HT showed a marked decrease in the caudate nucleus of brains of dementia subjects whilst 5-HIAA was unchanged. However, in the CSF 5-HIAA has been shown to be lowered and the probenecid loading test also showed lowered 5-HT turnover.

Recently there has been one report of a decrease in high affinity LSD binding in the temporal lobe, suggesting a decrease in 5-HT receptors, although it is not known whether they are pre- or postsynaptic (Bowen et al., 1979).

6.7 Brain monoamine enzymes

One problem which has been identified by Bowen (1980) is that several enzymes are sensitive to oxygen deprivation. Patients with Alzheimer's disease often die from bronchopneumonia. This problem was highlighted in the work of Bowen and colleagues (1976) who questioned their own earlier data on dopamine decarboxylase activity in brains from Alzheimer patients. Their later work suggested that the marked reduction which they had found in the activity of this enzyme in the brains of demented patients was in fact due to the oxygen deficit caused by bronchopneumonia.

Investigations of the amine degrading enzyme have been more revealing. MAO enzyme activity in the brain increases with age (Gottfries, 1980). In the brains of Alzheimer patients the activity of type B MAO (metabolizing phenylethylamine; see Section 2.4) is significantly higher than age matched controls. In contrast the activity of type A MAO (substrate: 5-HT) is unchanged. Gottfries (1980) suggests several explanations for this change including glial cell proliferation, relative to neuronal density. However the activity of platelet MAO (which is totally type B MAO) is also raised. This finding suggests that the disease causes a change in this enzyme occurring outside the central nervous system.

A recent study has shown decreased dopamine β-hydroxylase activity in the frontal and temporal cortex and hippocampus from post-mortem brains of Alzheimer patients, which was suggested to perhaps reflect changes in noradrenergic fibres in these regions (Cross et al., 1981).

6.8 Brain GABA

The early report of a decrease in glutamate decarboxylase (GAD) activity in post-mortem tissue for Alzheimer brains was not confirmed. The activity of the enzyme increases with increasing oxygen tension, and as stated above bronchopneumonia is a common terminal condition of dementia patients, which could account for the original finding.

One recent report has suggested that GABA receptors might be altered with GABA ligand binding being lower in some brain regions, indicating a receptor density decrease (Reisine *et al.*, 1978).

6.9 Brain cholinergic systems

There is now a considerable body of evidence to suggest that there are marked alterations in the cholinergic systems in the brains of Alzheimer patients, using tissue taken both at post-mortem and by biopsy (see Perry and Perry, 1980).

As mentioned in Section 2.16 choline is taken up into the brain by a high affinity uptake system and recently Sims *et al.* (1980) using biopsy material have provided evidence that this process may be reduced in brains of patients with Alzheimer's disease. Nevertheless CSF choline concentrations do not seem to be reduced in Alzheimer patients (see Fig. 6.2).

Choline acetyltransferase activity is decreased in the cortex, hippocampus, amygdaloid nucleus, mamillary body, and caudate nucleus. The decrease is most marked in those regions showing the most marked pathological change and are thus probably related to some extent with the morphological changes. Perry and Perry (1980) have shown elegantly the relationship between CAT activity and plaque formation in the neocortex (Fig. 6.3).

Similarly it has been shown that brains from Alzheimer patients have lowered activity of acetylcholinesterase (AChE) and there appears to be a relationship

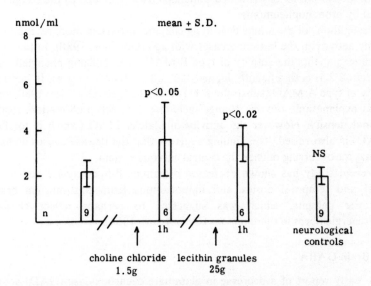

Fig. 6.2 Effect of oral administration of choline and lecithin on CSF levels of choline in patients with Alzheimer's disease. CSF value from fasting patients before and 1 h after, 1.5 g of choline chloride or 25 g lecithin. Neurological controls not fasted. (Reproduced from Yates *et al.* (1980) with permission of John Wiley & Sons Ltd)

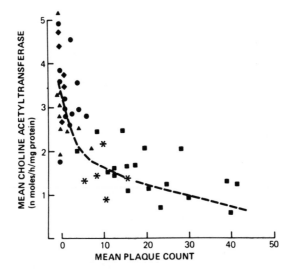

Fig. 6.3 Relation between choline acetyltransferase activity (CAT) and senile plaque formation in neocortex. Individual mean plaque counts plotted against mean cortical enzyme activities in 51 cases. ●, normal; ▲, depression; ♦, multi-infarct dementia; ■, Alzheimer's disease; *dementia of 'mixed pathology'. The correlation between CAT and plaque come in significant when calculated for the entire series. ($r = 0.82$, $P < 0.001$) and for those 23 cases with Alzheimer type pathology ($r = 0.45$, $P < 0.05$). (Reproduced from Perry and Perry (1980) with permission of John Wiley & Sons Ltd)

between the number of plaques and AChE activity although there does not appear to be any relationship between the degenerative changes and AChE activity.

In contrast the activity of butyrylcholinesterase (BChE) is increased in brains from Alzheimer patients and appears to be related to the neuropathological changes. In the hippocampus there is a good relationship between granulovacuolar degeneration and BChE activity (Fig. 6.4).

In general ligand binding studies have not shown any abnormality in the muscarinic receptor although there is one report of a decrease in receptor number in the hippocampus.

With regard to the acetylcholine precursor concentrations, choline has been reported to be raised both in plasma and CSF. However the large size of the precursor pool means that such data have little value.

Perry and Perry (1980) have presented preliminary observations which suggest that the activity of the two enzymes which produce the other precursor, acetyl CoA, namely pyruvate dehydrogenase and ATP-citrate lyase, are markedly reduced. Several other enzymes, including acetyl CoA synthetase were unchanged.

136

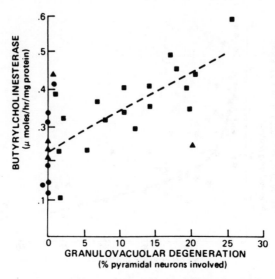

Fig. 6.4 Relation between butyrlcholinesterase ('BChE) and granulovacuolar degeneration in hippocampus. Individual enzyme activities are plotted against the percentage of hippocampal pyramidal neurons involved by granulovacuolar degeneration in 30 cases. ●, normal; ▲, depression; ■, Alzheimer's disease. The correlation is significant for the entire series ($r = 0.65$, $P < 0.001$) and for cases of Alzheimer's disease ($r = 0.68$, $P < 0.01$). (Reproduced from Perry and Perry (1980) with permission of John Wiley & Sons Ltd)

Perry and Perry (1980) have reviewed the data on abnormalities in brain cholinergic systems of Alzheimer patients and its possible interpretation and conclude that there may be a decrease in cholinergic function. If there is a change in cholinergic function it would appear to occur presynaptically since the study of Davies and Verth (1978) showed that muscarinic receptor number is not decreased.

PART 3   PHARMACOLOGY

6.10   Introduction

In general the drug treatment of the severely demented patient is limited and in terms of improvement there is little therapeutic success. The goals are those of control of disturbed behaviour, and maintenance of everyday living activities where possible. The major antipsychotic drugs are widely used for the former but the main treatment is socially based, simplifying tasks and helping patients adapt to their difficulties.

The major problem in proposing any rational pharmacological approach for dementia is the wide nature of the neurotransmitter defects that occur, together with the neuropathological changes. Parkinson's disease, in contrast, has defined pathological changes with relatively specific neurotransmitter changes allowing a rational approach to therapy (Chapter 8).

Some of the drug approaches used to treat demented patients are discussed below.

6.11   Improvement of cerebral blood flow and oxygenation

There have been several trials of hyperbaric oxygen; however, most trials were uncontrolled with poor patient selection criteria and recent well controlled studies have not indicated any value in this approach.

Two drugs have been used to try to improve cerebral blood flow: piracetam (1-acetamido-2-pyrrolidone) and hydergine which is a mixture of various ergot alkaloids. Controlled trials have not indicated any value of piracetam and the value of hydergine is so small as to be of little use. Indeed no drug at present available produces a reasonable increase in cerebral blood flow.

6.12   Cholinomimetic drugs

A major change occurring in Alzheimer patients is memory loss, and the cholinergic system has been shown to be involved in memory processes in animals and man (see the reviews of Deutsch, 1973; Karczmar, 1975; and Davis and Yamamura, 1978).

In man experiments in cognitive function have suggested that learning is impaired by the muscarinic cholinergic antagonist scopolamine and enhanced by cholinomimetic drugs such as choline (the ACh precursor) or arecoline.

Since cholinergic systems show major biochemical changes in the brains of Alzheimer patients (Section 6.8) several attempts have been made to increase their cholinergic function.

Choline is taken up into the brain by a high affinity uptake process (Section 2.16). Administration of choline chloride or lecithin (phosphatidyl choline) raises both plasma and CSF choline concentration (Fig. 6.2). Nevertheless the evidence that raising brain choline concentrations increases ACh function is a matter of some controversy. Furthermore after several clinical trials there is really no evidence that oral choline administration has any beneficial effect. Similarly the improvements obtained with lecithin have been minimal.

Recently another approach, that of using acetylcholinesterase inhibitors has been tried. Whilst it is still too early to evaluate fully, the work of Smith and Swash (1979) does suggest that physostigmine may have value in improving cognitive function in demented patients, particularly in reducing intrusion errors during recall experiments.

6.13 Future approaches

As stated earlier (Section 6.3) the widespread degenerative and neurotransmitter changes make it difficult to suggest rational therapeutic approaches. Indeed the changes seen are such that is is difficult to associate any one of them with any one psychiatric abnormality. It also means that it would be naïve to consider which are primary and which are secondary, particularly as old age produces its own neurochemical changes. The 'blanket' neurochemical investigations made at present are valuable but expensive and time consuming. Nevertheless the problems of dementia are large both to families and the community and it is to be hoped a rational pharmacological treatment will be found.

Chapter 7

Huntington's Chorea

PART 1 CLINICAL ASPECTS

7.1 Clinical features

This is a rare condition, with a clear genetic inheritance, of autosomal dominance with full penetrance. That is to say half of the children of an affected parent will inherit the disorder and be at risk of passing it on to their children and the other half will be free from the disease and not pass it on.

The condition usually expresses itself in middle age, and in the presence of a positive family history the diagnosis is easily made, although some cases apparently arise anew and the diagnosis here may be at first more difficult.

The earliest signs are unsteadiness and clumsiness with changes in personality. These develop gradually into choreiform movements and a generalized dementia.

The movements are continuous, jerky, and out of voluntary control. They may also be described as stretching, squirming, grimacing, grasping, and weaving, giving some idea of the generalization of the movement disorder and its severity. Difficulties with speech and gait are particularly troublesome consequences.

The dementia is similar clinically to those of other aetiologies. At the earlier stages, insight into the severe nature of the illness often causes depression, irritability, anger, and sometimes paranoid thinking. As the illness progresses these effects may be replaced by apathy and sometimes mild euphoria. Although slow in its course the illness progresses relentlessly and eventually profound dementia results.

Suicide occurs much more often than in the general population, and is also common in unaffected relatives.

There are no tests which unequivocally identify children who will be affected in later life, so 'genetic counselling' is difficult. In general, patients have had children by the time the disease manifests itself. One has either to advise all potential cases not to have families, knowing that this is unnecessary in half of them, or accept that the illness will continue in our society. This is a considerable ethical problem.

PART 2 BIOCHEMISTRY

7.2 Pathological changes

In Huntington's chorea there are major structural pathological changes. The basal ganglia show atrophy and the caudate and putamen can be markedly reduced in size. There is also some shrinking of the grey matter and these changes worsen with age. In the light of such observations and the fact that neurons are both fewer and often abnormal it is perhaps not surprising that there are major changes in neurotransmitter concentrations.

7.3 Brain GABA

It is now well established that GABA concentrations in specific regions of the post-mortem brain from patients are decreased. Perry, Hansen and Kloster (1973) reported a marked decrease in GABA in the striatum. The decrease was as much as 60% in substantia nigra and putamen–globus pallidus, a finding which was confirmed by Bird and Iversen (1974) who in addition examined occipital, frontal, and cerebellar cortex. A decreased GABA concentration was found only in the occipital cortex (Table 7.1).

Bird and colleagues (1973) also examined the activity of glutamic acid decarboxylase (GAD), the GABA synthetic enzyme in post-mortem choreic brain. Again major changes were found in caudate, putamen, and globus pallidus with activity decreased by approximately 75% (Table 7.2). This major enzymatic change has been confirmed by several other groups and again the predominant changes were seen in those areas which normally have the highest GABA concentration and GAD activity. No evidence has been obtained to suggest that these changes in any way result from the medication given to the choreic patients before death.

An initial report of a lowered GABA concentration in the CSF was not subsequently confirmed.

Ligand binding studies on post-mortem brains from patients did not reveal any marked change in GABA receptor sites (Enna *et al.*, 1976). This raises the possibility that some of the choreic movements might be alleviated by GABA agonists, although data so far have been disappointing (see Section 7.8).

Table 7.1 γ-Aminobutyric acid (GABA) in Huntington's choreic and control post-mortem brain (μ-mol/g wet wt)

Area	GABA Control $n = 13$	Huntington's chorea $n = 11$
Caudate nucleus	2.87 ± 0.22	1.57 ± 0.26[a]
Putamen/globus pallidus	4.45 ± 0.33	1.76 ± 0.24[a]
Substantia nigra	5.58 ± 0.28	2.18 ± 0.22[a]
Occipital cortex	2.01 ± 0.16	1.38 ± 0.12[b]
Frontal cortex	1.81 ± 0.16	1.59 ± 0.14
Cerebellar cortex	1.81 ± 0.14	1.61 ± 0.11

Different from control: [a] $P < 0.001$; [b] $P < 0.01$.

Adapted from Bird and Iversen (1977) with permission of John Wiley & Sons Ltd.

Table 7.2 Activity of glutamic acid decarboxylase in control and Huntington's choreic post-mortem brain ($\mu mol/^{14}CO_2$ evolved per h per g tissue; mean $\pm$ s.e.m)

Area	Control	Huntington's chorea	$P <$ (Student's test)
Globus pallidus	7.1 $\pm$ 1.3 (24)	1.9 $\pm$ 0.5 (14)	0.001
Substantia nigra	6.5 $\pm$ 1.0 (38)	2.2 $\pm$ 0.3 (40)	0.001
Griseum septum	6.4 $\pm$ 4.3 (4)	2.5 $\pm$ 1.7 (2)	N.S.
Caudate nucleus	5.1 $\pm$ 0.4 (68)	1.5 $\pm$ 0.2 (66)	0.001
Putamen	4.4 $\pm$ 0.4 (45)	0.9 $\pm$ 0.2 (41)	0.001
Dentate nucleus	4.3 $\pm$ 0.6 (21)	2.6 $\pm$ 0.6 (15)	0.025
Olfactory tubercle	3.1 $\pm$ 0.8 (11)	1.7 $\pm$ 0.3 (13)	N.S.
Hypothalamus	3.2 $\pm$ 0.5 (19)	4.4 $\pm$ 0.7 (18)	N.S.
Frontal cortex	3.0 $\pm$ 0.3 (30)	2.7 $\pm$ 0.3 (23)	N.S.
Hippocampus	1.9 $\pm$ 0.4 (15)	2.0 $\pm$ 0.4 (12)	N.S.

The number of brains are shown in parentheses.

Reproduced from Bird and Iversen (1977) with permission of John Wiley & Sons Ltd.

7.4 Brain catecholamines

Data on catecholamine changes in the brains of Huntington's chorea patients have not revealed the striking changes seen in the GABA system.

Bernheimer and Hornykiewicz (1973) reported a decrease in both DA and HVA concentration in the caudate of choreic brains but not in the putamen, whilst both Matsson (1974) and Bird and Iversen (1974) failed to detect significant changes. Subsequently Bird and Iversen (1977), looking at a larger group of brains, did detect a decrease in dopamine in the caudate but suggested that the change might be related to the degree of atrophy.

The fact that there is severe atrophy of the caudate nevertheless means that in terms of absolute amounts of dopamine in this region there must be a marked reduction of this monoamine and this is perhaps reflected in the reports of lowered CSF HVA in choreic patients.

In general there appear to be few changes in noradrenaline content of brains from choreic patients.

Both tyrosine hydroxylase and dopamine-β-hydroxylase activity have been examined in choreic brains but no major changes in either enzyme have been detected (Bird and Iversen, 1977).

7.5 Brain acetylcholine

Several groups have looked for changes in cholinergic systems in post-mortem brains from Huntington's patients.

Bird *et al.* (1973) reported a 50% reduction in the choline acetyltransferase activity of the basal ganglia of brains (Table 7.3) and this has been confirmed by other groups. The change correlated with the decreased GAD activity so it is possible that the decrease in the choline acetyltransferase activity is a consequence of the altered GAD function.

Use of ligand binding techniques has allowed the measurement of muscarinic acetylcholine receptor sites. Large decreases have been shown in the number of such sites in the putamen and caudate (e.g. Hiley and Bird, 1974) of brains from Huntington's patients. Since this change appears to be independent of the loss of choline acetyltransferase activity these receptors are possibly located on GABA neurons.

Table 7.3 Activity of choline acetyltransferase in control and Huntington's choreic post-mortem brain. Enzyme activity (μmol/h per g tissue; mean $\pm$ s.e.m)

Area	Control	Huntington's chorea	$P <$ (Student's test)
Putamen	21.8 ± 1.6 (41)	9.1 ± 1.3 (42)	0.001
Caudate nucleus	11.9 ± 0.8 (60)	5.4 ± 0.7 (64)	0.001
Olfactory tubercle	2.9 ± 0.4 (15)	2.5 ± 0.7 (15)	N.S.
Substantia nigra	1.2 ± 0.4 (12)	1.0 ± 0.5 (14)	N.S.
Frontal cortex	1.2 ± 0.1 (14)	1.4 ± 0.1 (11)	N.S.
Globus pallidus	0.9 ± 0.2 (9)	1.5 ± 0.3 (8)	N.S.
Hippocampus	0.3 ± 0.04 (16)	0.2 ± 0.04 (14)	N.S.
Hypothalamus	0.2 ± 0.02 (13)	0.2 ± 0.02 (11)	N.S.
Dentate nucleus	0.1 ± 0.02 (11)	0.1 ± 0.02 (7)	N.S.

The number of brains are shown in parentheses.
Reproduced from Bird and Iversen (1977) with permission of John Wiley & Sons Ltd.

7.6 Brain 5-HT

Neither brain nor CSF 5-hydroxyindole concentrations appear to be markedly changed in Huntington's chorea. However Enna *et al.* (1976) has reported a decrease in 5-HT receptor sites in the caudate. Again it is possible that these sites are located on neurons which have degenerated.

7.7 General conclusions

The data presented above give some explanation for the chorea observed in these Huntington's patients. This disease is, in some ways, considered to be the reverse of Parkinson's disease, being a hyperkinetic rather than an akinetic state. Biochemically too there are clear relationships; overactive dopamine systems (with the loss of inhibitory GABA systems) in the Huntington's patients. However, it is also apparent that whilst the biochemical findings provide some insight into the reasons for the symptomatology they give little assistance in the search for a primary defect. Furthermore, such studies have, to date, given only limited assistance in determining possible pharmacological approaches to treating this distressing disease.

PART 3 PHARMACOLOGY

7.8 Possible therapeutic approaches

The biochemical data reported above suggest three possible pharmacological approaches:

1. to increase GABA function;
2. to increase acetylcholine function;
3. to decrease dopamine function.

All these approaches have been tried.

Results with the first approach have been disappointing. Trials with baclofen (β-parachlorophenyl GABA), a possible GABA mimetic, and sodium valproate have been disappointing, (see Bird and Iversen, 1977) and whilst GABA transaminase inhibitors do exist, these drugs usually have several other actions as well.

Equally disappointing have been the trials with dimethylaminoethanol (deanol) which it is suggested is metabolized to acetylcholine. Use of cholinesterase inhibitors have also been unrewarding (see Bird and Iversen, 1977; Lloyd, 1978).

The major drugs used in Huntington's chorea are therefore the antidopaminergic drugs, tetrabenazine, phenothiazines, and butyrophenones. These drugs decrease the choreiform movements but clearly do not in any way reverse or even stop the progression of the disease. The pharmacology of these drugs was dealt with in Chapter 5.

Chapter 8

Parkinson's Disease

PART 1   CLINICAL ASPECTS

8.1   Historical aspects

The description by James Parkinson of the 'shaking palsy' or 'paralysis agitans' was published in an essay in 1817. Subsequently Charcot commented that the name 'paralysis agitans' was inaccurate and proposed the name Parkinson's disease. The term Parkinsonism is used to describe those patients who do not have idiopathic (unknown origin) Parkinson's disease but show the major features of Parkinson's disease, for example subjects who suffered from the epidemic of encephalitis lethargica and more recently the side effects of neuroleptic treatment (Section 5.15).

8.2   Clinical features

The predominant clinical signs are tremor, rigidity, akinesia, and the associated postural deformity. Other features can include ocular abnormalities, psychiatric disturbances, and alimentary disorders.

Tremor.   This is often the first complaint of patients and is seen in a high proportion of Parkinsonian patients. It is most marked in the hand, forearm, and foot, and occurs at rest, disappearing on voluntary movement. The head is not often affected, but the tongue and eyelids may be involved.

Rigidity, the akinetic syndrome, and postural deformity.   There is resistance to passive movement of the 'cog wheel' or intermittent type, slowness in executing voluntary movements, and few automatic movements. The rigidity may result in abnormal postures, especially of the fingers. Whilst rigidity and akinesia are associated, the relief of rigidity with stereotaxic surgery may still leave hypokinesia. It is probable that the mask-like face, reduced blinking, and speech problems are reflections of the rigidity. At worst the rigidity can be totally disabling, preventing the subject moving from a chair.

Psychiatric illness.   Parkinson did not include intellectual changes in his original classification, but later workers have found dementia to be common. Recent studies have suggested a prevalence of up to 30%, with an association with akinesia, and perhaps unsurprisingly, age. This rate is of course much higher than that of an age matched control group.

Depression is even more common, but this may be secondary to the consequences of having the illness. Treatment with L-dopa even when successful neurologically is said to increase the risk of depression.

The precise nature of the relationship with psychiatric disorders remains to be elucidated.

8.3 Prevalence

Overall the prevalence of Parkinson's disease in the United Kingdom is around one per thousand, although the fact that it is a disease of later life puts the rate at approximately one per two hundred for those over 50 years old. Marsden (1976) reports that there are between 60,000 and 80,000 sufferers in the UK with around 15,000 in hospital care, 22,000 handicapped in the community, and 30,000 not severely handicapped.

Postencephalitic Parkinsonism is now of decreasing importance, but the epidemic of encephalitis in the early 1920s affected over one million people, many of whom developed Parkinsonism. The prevalence of drug induced Parkinsonism is more difficult to assess, but there are probably some tens of thousands of patients receiving anti-Parkinsonian drugs in the UK at the present time.

8.4 Natural history

Idiopathic (unknown origin) Parkinson's disease is a progressive disease. Some patients die within a year but some survive 30 years. The mean survival time is approximately 9 years and the mortality for patients with Parkinson's disease is three times that of an age matched population. L-Dopa therapy (Sections 8.12 and 8.13) will improve the clinical state of many patients but does not markedly alter the natural history.

8.5 The 'on–off' phenomenon

This is a well recognized feature of Parkinson's disease. In the USA it is often referred to as a 'swinging'. It describes fluctuations in performance but is complicated by the fact that it is used to describe both a pathological state and a drug induced effect.

Prior to the advent of L-dopa therapy fluctuations had been well described. Freezing could occur during periods of more normal activity and stress could worsen the symptoms. Following the introduction of L-dopa it was seen that the pathological on–off symptoms subsided, however after 2–3 years on the drug the on–off phenomenon returns. It is not clear why there is this delay in the phenomenon recurring; however, appearance of the on–off effect seems to coincide with the plasma levels of the L-dopa. Following a dose, improvement occurs, but after 2–3 hours when the plasma L-dopa concentration falls, the Parkinsonian state returns, until a further dose is given. This effect worsens with time. Further, the on–off phenomena can include the severe dyskinesia that may occur with the peak of the L-dopa effect, that is 2 hours after administration.

At worst the on–off effect can be disastrous for the patient who experiences a rapid (in minutes) swing from mobility and gross dyskinesia to akinesia and total immobility. It remains a major problem of Parkinson's disease and L-dopa therapy.

PART 2   BIOCHEMISTRY

8.6   Introduction

Parkinsonism may be drug induced (Section 5.15). However in Parkinson's disease itself there are clear neurochemical and pathological changes.

8.7   Dopamine in the substantia nigra

The first major neurochemical change reported was by Ehringer and Hornykiewicz (1960) in a paper now rightly regarded as a classic in neurology and which can be said to have led directly to the major treatment now used. In this paper and several subsequent studies Hornykiewicz and his colleagues showed that there is a severe deficit in dopamine and its metabolites in the nigrostriatal system.

Table 8.1 taken from Hornykiewicz (1973) shows that in the substantia nigra

Table 8.1   DA and HVA in some nuclei of the basal ganglia in patients with Parkinson's disease

Brain region	Controls	Parkinsonism (not classified)	Idiopathic	Parkinsonism Postenke- phalitic	'Arterio- sclerotic'
Caudate nucleus					
DA	2.64 (28)	0.36 (7)	0.40 (5)	0.06 (4)	0.63 (6)
HVA	3.23 (6)	0.76 (7)	1.10 (5)	0.50 (4)	1.28 (6)
Putamen					
DA	3.44 (28)	0.19 (7)	<0.03 (5)	0.05 (4)	0.20 (6)
HVA	4.29 (6)	0.79 (7)	1.07 (5)	0.58 (4)	1.28 (6)
Substantia nigra					
DA	0.46 (13)	0.07 (10)	—	—	—
HVA	2.32 (7)	0.41 (9)	—	—	—
Globus pallidus					
DA	0.30 (8)	0.14 (6)	—	—	—
HVA	2.25 (8)	0.72 (9)	0.63 (4)	0.57 (4)	0.81 (5)

Reproduced from Hornykiewicz (1973) by permission of the Medical Department, The British Council.

(Figures are mean values in µg/g fresh tissue; numbers of cases examined are in parentheses)

and the terminal regions of the striatal pallidal regions dopamine and the major metabolite HVA are markedly decreased. This is seen in post-mortem brains obtained from subjects who suffered from idiopathic, postencephalitic, and arteriosclerotic Parkinsonism.

In monkeys Poirier and Sourkes (1965) showed that electrolytic lesions of the substantia nigra led to marked decreases in the concentrations of DA and HVA in the striatal regions. Lesions of this type will also lead to marked loss of the dopamine synthesizing enzymes, tyrosine hydroxylase and dopa decarboxylase. Exactly the same loss of enzyme activity is found in the striatal regions of brains from patients with Parkinson's disease.

There is a reasonable correlation between the amount of cell body loss in the substantia nigra and the dopamine loss in the striatum.

The dopamine terminal loss was also shown by the observation that MAO inhibitor administration before death did not lead to an increase in brain dopamine concentrations. In contrast both NA and 5-HT showed marked increases, even though these transmitters were lowered in the brains of Parkinson patients not treated with the drug.

Hornykiewicz (1966) has suggested that the release of the inhibitory control of the dopaminergic pathway from the substantia nigra to the pallidus is responsible for the rigidity of Parkinsonism whilst the akinesia is due to lack of dopaminergic control in the neostriatum. Certainly lesions of the nigrostriatal pathway of monkeys produce hypokinesia and there is a correlation between the CSF HVA decrease and severity of akinesia in Parkinsonian patients. Furthermore L-dopa administration (Secion 8.2) relieves the akinesia.

In contrast there is no close temporal relationship between loss of tremor and L-dopa administration, nor do nigrostriatal pathway lesions produce tremor.

8.8 Cerebrospinal fluid dopamine metabolite concentrations

There have been several reports of a lowering of HVA in the CSF of Parkinsonian patients. Not surprisingly the rate of HVA accumulation following probenecid is also lowered, indicating a lowered rate of dopamine synthesis. It has been suggested that a correlation exists between the probenecid induced rise of HVA and the severity of the rigidity.

8.9 Brain 5-HT metabolism

The early studies of Hornykiewicz and colleagues also found some lowering of 5-HT concentrations in caudate, putamen, pallidus and thalamus of Parkinsonian brains. Administration of a MAO inhibitor increased the brain 5-HT concentrations markedly.

It has been reported that CSF 5-HIAA is lowered in Parkinsonian patients and that the rate of accumulation following probenecid is somewhat diminished.

Despite these biochemical indications of an apparent 5-HT involvement in Parkinsonism, pharmacological studies have failed to provide evidence for the changes in 5-HT having a major role in the pathology of Parkinson's disease and it is generally felt that the 5-HT changes are secondary to the major changes in dopamine function.

8.10 Brain GABA

The GABA concentration in the caudate nucleus and putamen of Parkinsonian patients does not seem to be decreased. However, there is at least a 50% loss of glutamic acid decarboxylase (GAD) activity in both these regions and the substantia nigra of Parkinsonian patients (Lloyd, 1980).

GABA receptor number is also decreased in the substantia nigra but not the corpus striatum or cerebral cortex (Table 8.2).

Table 8.2 Specific binding of [^{3}H]-GABA to membranes prepared from different brain regions of control of Parkinsonian patients

Brain region	[^{3}H]-GABA binding (fmol/mg protein)		
	Control	PD	% Control
Caudate nucleus	66.3 ± 8.2 (19)	70.6 ± 9.4 (6)	107
Putamen	67.8 ± 13.7 (17)	91.7 ± 19.6 (8)	135
Substantia nigra	30.8 ± 5.0 (11)	9.7 ± 2.9 (6)	31*
Temporal cortex	187.8 ± 27.6 (12)	240.8 ± 14.8 (5)	129

Data expressed as mean ± s.e.m. Number of patients studied in parentheses.
*$P < 0.01$ vs control patients.
Adapted from Lloyd (1980) with permission of John Wiley & Sons Ltd.

Lloyd (1980) has suggested that these receptors are probably located on the cell bodies or dendrites of the dopamine neurons originating in the pars compacta. GABA is normally an inhibitory transmitter and thus GABA mediated tonic inhibition of remaining nigrostriatal dopamine neurons is probably reduced.

In Parkinson's disease there is evidence that the intact nigrostriatal neurons are maximally activated and it has been proposed that this is due, in part, to disinhibition of the striatonigral GABA input. This is supported by the observation that GAD activity increases with time on L-dopa treatment (Fig. 8.1), suggesting that trans-synaptic changes have occurred and GABA function has been increased. Increasing dopamine function with L-dopa is apparently increasing GABA function.

The importance of this may be in therapeutic approaches and is discussed in Section 8.23.

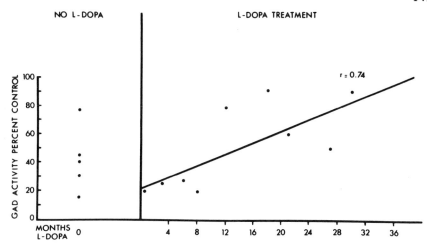

Fig. 8.1 Glutamic acid decarboxylase (GAD) activity in the putamen of patients with Parkinson's disease. (Reproduced from Lloyd and Davidson (1979) with permission of Raven Press)

PART 3 PHARMACOLOGY

8.11 Introduction

The observation of Ehringer and Hornykiewicz (1960) that there is a marked decrease in the concentration of dopamine in the corpus striatum suggested a logical pharmacological approach to treatment, that of trying to increase brain dopamine concentrations. This has been the basis of almost all the modern therapy of Parkinsonism. The fact that most dopamine antagonists produce Parkinsonism (Section 5.15) was a further indication that the major clinical features of Parkinsonism result from a deficiency in dopaminergic activity in the brain.

Before the advent of L-dopa therapy the main therapeutic approach had been to use anticholinergic drugs. Charcot in 1867 had advised the use of the antimuscarinic drug atropine and for the next century atropine and related drugs had formed the basis of the pharmacological treatment. In addition stereotaxic surgery was sometimes used, producing lesions in the ventrolateral nucleus of the thalamus. The effect of this was relief of tremor and rigidity in the contralateral limbs. Bilateral surgery often resulted in a much increased incidence of complications and was therefore less used.

8.12 L-Dopa therapy

Dopamine does not cross the blood–brain barrier and cannot be given to increase brain dopamine concentrations; and administration of loading doses of tyrosine does not markedly alter the rate of dopamine formation, presumably

Table 8.3 Chronic oral L-dopa therapy (2–6 g daily) in patients with Parkinson's disease: effect on brain levels of dopamine, homovanillic acid, dopa and 3-O-methyldopa

Brain region	Dopamine	Homovanillic acid	Dopa	3-O-Methyl-dopa
Caudate nucleus				
Non-dopa treated	0.47 ± 0.33	1.18 ± 0.10	nd*	nd
	(3)	(3)	(3)	(3)
Dopa treated	2.22 ± 0.93	8.16 ± 3.10	0.24 ± 0.14	1.66
	(4)	(4)	(4)	(4)
Putamen				
Non-dopa treated	0.26 ± 0.09	0.67 ± 0.27	nd	nd
	(3)	(3)	(3)	(3)
Dopa treated	2.06 ± 0.71	11.46 ± 3.28	0.25 ± 0.18	3.00 ± 2.43
	(4)	(4)	(4)	(4)
Temporal cortex				
Non-dopa treated	nd	nd	nd	nd
	(2)	(2)	(2)	(2)
Dopa treated	nd	3.01 ± 0.92	1.01 ± 0.54	3.28 ± 2.56
	(3)	(4)	(3)	(3)

*nd, not detectable.

Reproduced from Hornykiewicz (1973) by permission of the Medical Department, The British Council.

(Figures are mean values in µg/g fresh tissue ± s.e.m.; numbers of cases examined are in parentheses)

because of the rate-limiting nature of tyrosine hydroxylase (Section 2.8). In contrast L-dopa administration to experimental animals produces significant increases in cerebral dopamine concentrations, while having little effect on brain noradrenaline content. The high activity of the decarboxylase enzyme ensures a rapid increase in dopamine formation.

Following the initial report of decreased dopamine content in the nigrostriatal system, two independent trials of the efficacy of L-dopa therapy were undertaken in Vienna and Montreal. Whilst the results were encouraging, using doses up to about 1 g, the improvement was limited and there were various side effects such as nausea, which limited its usefulness.

Cotzias, Van Woert and Schiffer (1967) used much higher doses (6–8 g) and reported that if the dose was stepped up slowly these large doses could be tolerated. Furthermore the clinical improvement was now marked, with some patients showing quite striking recoveries. Nevertheless side effects such as nausea, vomiting, and hypotension still occurred.

Administration of large doses of L-dopa does markedly increase the concentration of dopamine and its metabolites in the brain of patients suffering from Parkinson's disease (Table 8.3).

8.13 Use of peripheral decarboxylase inhibitors with L-dopa

The side effects of L-dopa were sometimes sufficiently troublesome to make treatment impossible. It was suggested by Bartholini and his coworkers that the side effects might have been the result of catecholamine formation in the periphery. The decarboxylase enzyme is widely distributed in gut and blood vessels and much of the L-dopa administered is decarboxylated peripherally. To achieve a sufficient increase in cerebral L-dopa concentration therefore high doses have to be given. Bartholini and Pletscher (1967) suggested the possibility of improving therapy, and decreasing side effects by administration of a decarboxylase inhibitor which would not cross the blood–brain barrier.

The approach used by the drug companies concerned was elegant in its simplicity. Merck, Sharp, and Dohme, for example, took the established dopa decarboxylase inhibitor α-methyldopa and added a hydrazine group (producing α-methyldopa hydrazine). The hydrazine grouping prevented the drug passing the blood–brain barrier but did not prevent the decarboxylase inhibitory action of the drug. Roche developed seryltrihydroxybenzylhydrazine to the same end.

The result of combining L-dopa with a peripheral decarboxylase inhibitor was to lower the incidence of side effects, particularly nausea and to lower the therapeutically useful dose of L-dopa to 3–4 g daily (Cotzias, Papavasiliov and Gellene, 1969). In general the preparations of L-dopa now available are combined tablets of L-dopa with a peripheral decarboxylase inhibitor.

Use of such preparations does not prevent the L-dopa induced dyskinesias, which are a consequence of the action of L-dopa in the brain. However, lowering the dose will generally lead to their disappearance.

8.14   L-Dopa plus pyridoxine administration

The administration of pyridoxine was originally thought to be useful in enhancing the decarboxylase step since the enzyme is pyridoxal phosphate dependent. However, it was found that the therapeutic effects of L-dopa were lessened. This is almost certainly due to the formation of a Schiff's base by the reaction of L-dopa with pyridoxal-5-phosphate, thereby leading to an inactive compound. Pyridoxine is now considered to be contra-indicated in L-dopa therapy.

8.15   L-Dopa plus a MAO inhibitor

In attempts to increase the therapeutic effect of L-dopa, early trials in Vienna combined the amino acid with MAO inhibitors. The side effects produced were severe, however, and the problems with MAO inhibitors and food reactions (Section 3.22) together with the efficacy achieved with L-dopa with a peripheral decarboxylase inhibitor meant that this drug combination was not pursued.

Recently a new MAO inhibitor has been available, the selective MAO type B inhibitor (Section 2.4) deprenil. Being a MAO-B inhibitor only, it does not produce the hypertensive crisis of the older non-selective inhibitors. Furthermore it has been suggested that dopamine is predominantly metabolized in human brain by the type B enzyme (Glover et al., 1977). Certainly post-mortem brains from Parkinsonian patients show greater inhibition of MAO activity towards DA than 5-HT when the patients have been on deprenil. Several trials have demonstrated the usefulness of combining deprenil with L-dopa (see for example Birkmayer *et al.*, 1977), but further trials are needed as there is still some controversy as to the efficacy. The trials did suggest that deprenil significantly reduced the on–off phenomenon (Section 8.5) and the appearance of the severe akinetic states. If we assume that the on–off state is due in part to the fluctuations in the amount of dopamine being formed, it seems reasonable to suppose that deprenil is acting by inhibiting transmitter metabolism and increasing the time that the dopamine is present. However, since a metabolite of deprenil is amphetamine it may be also that this drug enhances dopamine release.

When deprenil is combined with L-dopa the dose of L-dopa can be lowered.

8.16 The value of L-dopa

Whilst not everyone benefits from L-dopa therapy, its value must not be underestimated. Marsden (1976) estimated that approximately one-fifth of patients will not show a response, one-fifth modest improvement, two-fifths moderate improvement, and one-fifth spectacular improvement. In the light of the near total failure in treating this condition before the advent of L-dopa such figures are remarkable.

Nevertheless it is important to remember that Parkinson's disease is a degenerative disease and L-dopa does not alter the progress of the disease but affects only the symptoms.

There are some data suggesting that those patients most helped by L-dopa show the greatest rise in CSF HVA on L-dopa therapy. Presumably these are the patients with the greatest number of intact terminals and thus able to metabolize L-dopa most effectively. Nevertheless against this it should be remembered that there appears to be little relationship between severity and improvement on L-dopa.

There must be a marked loss of dopamine (probably more than 50%) for Parkinson's disease to be manifested. One of the reasons that L-dopa works in patients with marked loss of dopamine neurons is that denervation supersensitivity almost certainly occurs and this results from an increase in dopamine postsynaptic receptor number (Creese, Burt and Snyder, 1977). It is also probable that L-dopa is decarboxylated in other terminals (5-HT) and cerebral small blood vessels and dopamine so formed arrives in the area diffusely in a 'cloud' rather than merely as the result of decarboxylation in specific dopamine nerves.

8.17 Bromocriptine

Since degeneration continues during L-dopa therapy an obvious alternative therapeutic approach was the use of dopamine agonists, since such drugs would not require the metabolic step necessary in L-dopa therapy. One such agonist tried was bromocriptine, an ergot compound. It is a complicated drug insofar as it behaves as a dopamine agonist in some animal tests but apparently is most effective in the presence of intact presynaptic dopamine stores.

Bromocriptine is undoubtedly effective in Parkinsonism. In experimental animals it can be shown to have a slow onset of action, but then produces dopamine agonist actions over several hours. This long lasting action is reflected in its use in Parkinson's disease. In patients who tend to have a marked drug induced on–off effect this long lasting action of bromocriptine can make it the drug of choice since it lessens the on–off effect. On the other hand, compared with L-dopa it is an expensive drug and therefore not in general the first choice.

It was thought that some of the side effects of L-dopa therapy might be due to the formation of various dopa metabolites. The fact that bromocriptine produces many of the same effects indicates that these unwanted effects occur as the result of dopamine mimetic activity.

8.18 Future approaches to increase dopamine function

Effort is currently being expended to produce more selective dopamine agonists based on the observations that there is more than one type of dopamine receptor in the brain (Kebabian and Calne, 1979). Such approaches have yet to be examined clinically.

8.19 Amantadine

This drug was introduced and is used as an antiviral drug; however, it is also useful in treating Parkinsonism. In mild Parkinson's disease it is useful, being better than the anticholinergics but not as effective as L-dopa. It is of particular value in reducing the akinetic state. It is also relatively free from side effects. The drug is not anticholinergic nor a dopamine agonist but animal studies have indicated that it releases dopamine from terminals. In particular the effects of amantadine and L-dopa have been shown to be additive.

8.20 Anticholinergics

There is a probable functional 'balance' between dopamine and acetylcholine in the striatum. Dopamine antagonists generally produce, at least initially, Parkinsonian side effects, but this is minimized when the drugs also have anticholinergic effects (see Section 5.18). Tremor is also a feature of cholinergic drugs such as oxotremorine.

In a similar way mild Parkinson's disease, where there is some loss of dopamine function and mild tremor, can be alleviated by administration of anticholinergics. The most commonly used drugs are benzhexol or orphenadrine. These are administered alone or sometimes with L-dopa.

8.21 Electroconvulsive therapy

There are several anecdotal reports that in patients with depression and Parkinson's disease, treatment of the depression with ECT resulted in a marked improvement in the Parkinsonian state. Recently Balldin and his colleagues

(1980) on the basis of the animal studies showing that electroconvulsive shock resulted in enhanced postsynaptic dopamine responses (see Section 3.29) undertook an open trial of ECT in Parkinson's disease. The results were encouraging, particularly in the alleviations of the on–off phenomenon. However, further trials are clearly necessary before proper evaluation can be made.

8.22 PLG

Proline–leucine–glycine amide is a tripeptide that has been under consideration as a possible neurotransmitter (see Section 2.36). It has been shown to potentiate the behavioural actions of a MAO inhibitor and L-dopa in experimental animals. Recently Barbeau (1979) has reported that an intravenous bolus of PLG produced a short lasting improvement in motor performance. Rigidity and tremor were improved.

The use of peptides to treat Parkinsonism is novel but awaits further careful controlled studies.

8.23 GABA mimetics

It was suggested in Section 8.10 that L-dopa therapy in turn increases GABA function and alters a dopamine–GABA balance. Certainly in animals the cataleptic effect of neuroleptics can be alleviated by inhibiting GABA function, and it was felt therefore that a GABA mimetic might alleviate the L-dopa induced dyskinesia. Whilst one drug has been tried, it appears that it may alleviate the dyskinesias but aggravate the underlying Parkinson's disease (Lloyd, 1980). Such approaches therefore are unlikely to be clinically useful.

8.24 Conclusion

The euphoria that occurred in neurology following the therapeutic introduction of L-dopa has now disappeared. It was one of the first cases of the use of pharmacology to treat a major neurological disease, the approach was logical, based on neurochemical data, and the clinical results were sometimes astonishingly good.

It is now recognized that L-dopa therapy can produce marked problems, such as dyskinesias and these have often to be 'balanced' with akinetic state if the drug is not given, in deciding how the drug is to be given. The drug does not prevent the underlying degenerative changes, the aetiology of which remains obscure.

In general the feeling now is that L-dopa should not be given in the early stages of the disease, despite the improvement it can bring, but should be introduced as the disease progresses.

Chapter 9

Drug Dependency

9.1 General introduction: social aspects, tolerance, and dependence

Although social and medical attitudes to non-therapeutic drug use have varied markedly with time and place, the use of substances which affect the mind and alter aspects of consciousness has been recognized in almost all societies, and at all times.

In Western society today alcohol and nicotine have acknowledged major roles in many social interactions, and similarly with cannabis in some subcultures. It is clear that it is inadequate to consider drug dependency in narrow medical or pharmacological terms, but that social and psychological factors demand consideration.

The most recent World Health Organization definitions of drug abuse and dependence are seen to go some way towards recognizing this.

1. A drug is any substance that, when taken into the living organism, may modify one or more of its functions.
2. Drug abuse is persistent or sporadic excessive use of a drug inconsistent with, or unrelated to, acceptable medical practice.
3. Drug dependence is a state – psychic and sometimes also physical – resulting from the interaction between a living organism and a drug, characterized by behavioural and other responses that always include a compulsion to take the drug on a continuous or periodic basis in order to experience its psychic effects, and sometimes to avoid the discomfort of its absence. Tolerance may or may not be present. A person may be dependent on more than one drug.

The various aspects of dependence, tolerance, physical and psychic dependence, and withdrawal states are further defined.

Tolerance is the phenomenon of the need to increase the dose of drug in order to maintain the effect. Several mechanisms may be responsible for this although it should be emphasized that the mechanisms of tolerance are poorly understood in pharmacological terms.

Tolerance can result from induction of the cytochrome P450 drug metabolizing enzyme systems. This leads to a more rapid clearance of the active drug and therefore a lessening of its effect. Cross-tolerance refers to the phenomenon of use of one drug producing tolerance to another drug. Since several abused drugs are metabolized by the drug metabolizing enzyme system, induction of the enzymes by one drug leads to tolerance of other compounds. A good example of this is the increased dose of anaesthetic required in heavy alcohol users.

A second possible reason for tolerance developing is receptor desensitization. Some abused drugs, opiates (Section 2.28), for example, appear to produce their mood altering effects by interacting with a known receptor system. There is therefore an increase in receptor stimulation. This will lead to receptor sub-sensitivity and consequently a higher concentration of compound will be necessary to produce the same effect. Cross-tolerance can be involved in this mechanism also since several hallucinogenic compounds stimulate the same receptors (Section 9.14) so desensitization of the receptor will result in lowered responses to other drugs acting at the same site.

A further possible reason for tolerance could be the endogenous production of antibodies to the abused drug thereby lessening its effect. Such a phenomenon has been observed with the phenothiazine drugs and could well occur following prolonged use of other drugs.

Physical dependence is an adaptive state which is manifest by physical disturbances when the administration of the drug is suspended or when its action is opposed by the administration of a specific antagonist (withdrawal states). *Psychic dependence* is the intense craving and compulsive perpetuation of abuse, to repeat the desired effect of the drug.

The distinction between physical and psychic dependence is in practice difficult to maintain, and is sometimes difficult to justify.

Although these definitions emphasize the common features of dependency, there are major differences in the effects of different drugs and consequently their use, reflecting largely their pharmacological properties.

It is also important to recognize that drug abuse is often not restricted to a single substance.

PART 1   ALCOHOL

9.2   Social aspects

Alcohol is a simple chemical formed naturally from the fermentation of sugars by yeasts. It is therefore easy to produce and is extensively used.

It is unique in this chapter in being the only drug with prominent central nervous system effects which is legally available without medical prescription. It is consumed in enormous quantities throughout the world but the ease of its availability varies from country to country. Availability also varies with time, as legislators attempt to reduce the problems associated with overconsumption by making it either easier or more difficult to obtain, through licensing and restrictions of sales. In contrast to other drugs alcohol is totally proscribed by countries on religious rather than social grounds.

Most authorities are now agreed, however, that the primary factor in a country's total consumption is *price*, and whilst some individuals will sacrifice other pleasures for the sake of alcohol, overall an increase in price causes a reduction in alcohol related disabilities.

9.3 Physical effects

The effects of alcohol are familiar. It is a central nervous system depressant and the apparent stimulation arises from release of inhibitory systems. At low doses (50 mg/100 ml blood) there is a sense of well being, reduction in anxiety, and increase in sociability and talkativeness.

At levels of around 100 mg/100 ml blood there are marked changes in mood which are variable, including depression and aggression. At this level there is also a considerable degree of clumsiness and motor incoordination.

A higher levels (300 mg/100 ml) most individuals would be regarded as 'drunk' with confusion, unsteadiness, slurred speech, and progressively including hypotension, respiratory depression, coma, and death.

It is worth emphasizing that performance is demonstrably impaired by even small amounts of drink, but paradoxically, drinkers believe their performance to be improved. Any legal limits set for proof of intoxication are therefore arbitrary.

There is marked tolerance to the effects of alcohol, with hardened drinkers able to consume vastly greater amounts of alcohol than novices. Far from being evidence of masculinity this is relatively easily achieved by enzyme induction and CNS adaptation (Section 9.1).

Withdrawal phenomena are experienced after even a single excessive intake. Headache, nausea and vomiting, tremor and sweating occur, with in more severe cases hallucinations, most commonly visual and tactile, and convulsions. These are abolished by renewed intake of alcohol, which therefore reinforces continued consumption.

9.4 Clinical aspects

Attitudes towards the overconsumption of alcohol have changed over this century. Initially regarded as a moral vice, it was later seen as the illness of alcoholism. This change had many advantages in terms of treatment and social acceptability, but has more recently itself been criticized on several grounds.

Original studies on alcohol abuse were biased by patient selection, giving a false idea of the nature of dependence, and missing many forms of alcohol abuse. Further the use of a label such as alcoholism implies a distinct condition, but attempts to define the term have been less than successful.

It is now regarded as being more useful to distinguish between an alcohol dependence syndrome, and alcohol-related disabilities, which may occur independently or coexist. The former refers to the dependence on alcohol use, with socially abnormal drinking, tolerance, and withdrawal symptoms, and the latter to the physical, mental, or social consequences of the use of alcohol. These consequences are wide ranging including family and work difficulties, accidents, crime, personality changes, intellectual deterioration, suicide, liver damage, cancer, heart

158

disease, and infections. In addition there is an association with depression, mania, and schizophrenia. Readers are referred to standard texts for a fuller description of the clinical aspects.

Treatment of those who desire it is often difficult, and includes a wide range of approaches including social, psychotherapeutic, behavioural, and pharmacological. Outcome is assessed in a variety of ways including social and physical functioning. The old idea that the only satisfactory outcome in terms of alcohol consumption is total abstinence is now known to be false, and a return to 'social' drinking is not unusual.

9.5 Metabolism

Pharmacologically the effects of alcohol are complex. Alcohol is rapidly absorbed and distributed and is then metabolized to the aldehyde and acid (Fig. 9.1).

Alcohol dehydrogenase is a widespread enzyme, although it is found predominantly in the liver. It is an enzyme requiring NAD as a cofactor. The second enzyme is aldehyde dehydrogenase. This is also a NAD-requiring enzyme and is a relatively non-specific aldehyde metabolizing enzyme, again predominantly though not exclusively present in liver. Aldehyde concentrations rise after alcohol administration suggesting that this step is rate limiting. The enzyme is inhibited by disulfiram, a compound which also inhibits dopamine-β-hydroxylase. Disulfiram is used in the treatment of alcoholism. On its own disulfiram has little effect. However ingestion of alcohol by patients taking it results in an extremely unpleasant syndrome of nausea, vomiting, flushing, and sweating. This has the effect of discouraging further alcohol consumption. Normally it is used together with other supportive measures.

Repeated heavy alcohol use has the effect of inducing its own metabolism, resulting in tolerance. The metabolism of other drugs is also increased. Common examples include the barbiturates and various anaesthetics, the pharmacological action of both being decreased in alcoholics.

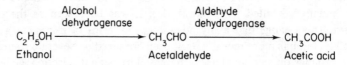

Fig. 9.1   The major metabolic pathway of alcohol (ethanol)

9.6 Mechanism of addiction

The neurochemical effects of alcohol have been extensively examined in the hope of elucidating the addictive effects of the drug.

Many studies have concentrated on the effects of alcohol on the existing neurotransmitter systems, particularly 5-HT and the catecholamines. Data have been conflicting, probably in part as the result of the many different experimental protocols. However, there have been some fairly consistent observations. Ethanol enhances the accumulation of labelled catecholamines following injection of radiolabelled tyrosine, indicating increased synthesis of both dopamine and noradrenaline. Recently it has been demonstrated that in chronic alcohol treated rats there is an increase in postsynaptic dopamine-mediated behavioural responses in the n. accumbens (Liljequist, 1978). With regard to 5-HT it has been found that ethanol administration increases the rate of 5-hydroxytryptophol formation from 5-hydroxyindoleacetaldehyde to the detriment of the 5-hydroxyindoleacetic acid pathway. However, whilst this change certainly occurs in the liver a similar shift does not apparently occur in the brain (see Green and Grahame-Smith, 1975). In contrast, no consistent changes in GABA metabolism have been seen although there have been some suggestions that altering GABA function can interfere with ethanol-mediated behaviours.

Recently there has been much interest in the possible interaction of acetaldehydes with the monoamines. Under suitable conditions the monoamines can react with acetealdehyde to form β-carbolines or tetraisoquinolines. The reaction is the well known Pictet–Spengler reaction, illustrated in Fig. 9.2. These compounds have marked pharmacological actions. Of particular interest are the studies on alcohol preference. Myers and his colleagues have shown that intraventricularly infused tetraisoquinolines result in rats preferentially consuming alcohol. Other studies have reported both confirming and conflicting results and it may be that the variation in techniques used result in the apparent controversies. This area has recently been reviewed (Deitrich and Erwin, 1980). This review also includes a summary of the many pharmacological actions of the β-carbolines and tetraisoquinolines.

In view of the fact that seizures that can occur during the alcohol abstinence syndrome it is also worth noting that β-carbolines may interact at the benzodiazepine receptor site (see Section 4.7) and change seizure threshold.

It should also be noted that β-carbolines can be hallucinogenic (e.g. the harmala alkaloids) and tremorogenic.

Of course the foregoing only becomes of importance if these compounds can be formed in the brain, and there is some evidence for this. 5-Methyltetrahydrofolate will react under physiological conditions with 5-HT or tryptamine to produce a β-carboline, although the reaction is probably not enzymatic (Fig. 9.2). Furthermore administration of acetaldehyde to rats will increase the fluorescence in the brain which has tentatively been ascribed to the formation of β-carbolines. The papers

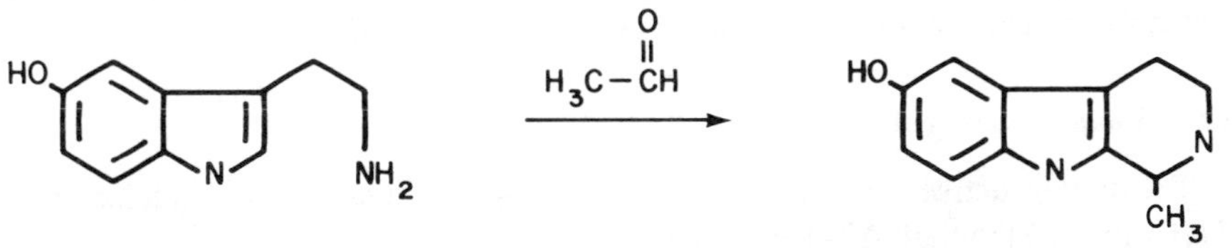

Fig. 9.2 Pictet–Spengler reaction. Figure shows the reaction of 5-HT with acetaldehyde to form a β-carboline

of Deitrich and Erwin (1980) and Elliot and Holman (1977) review this area well. A good review of the possible involvement of acetaldehyde in alcoholism has also been published (Lindros and Eriksson, 1975). It is apparent from the reviews cited above that this area is one of much speculation and little hard evidence. One of the current difficulties is that of specific analytical techniques. Nevertheless it is an area of great promise. Whilst the β-carbolines and tetraisoquinolines may explain some of the behavioural and physical changes associated with alcohol abuse the mechanisms involved in the addictive process are still unclear. It is also quite possible that these compounds are involved in disease states quite unrelated to alcoholism.

PART 2   CANNABIS OR MARIJUANA

9.7   Physical and psychological effects

Cannabis vies with alcohol for the distinction of being the most controversial drug of abuse, with a similar diversity of laws governing its use.

It is derived from the hemp plant *Cannabis sativa*, which grows freely and has been cultivated to make ropes and bird seed.

It is used mostly in the form of the dried plant (marijuana) especially the flowering tops and growing leaves, or as resin (hashish) but very occasionally also as the purified substance (see below).

Slang names include, bhang, dope, grass, joint, Mary Jane, pot, reefer, roach, shit, and weed.

Cannabis is either smoked, with or without tobacco, or ingested in food or drink. The effects come on rapidly (2–3 minutes) with the former and more slowly with the latter (1 hour) and last for about 6 hours.

The physical effects include tachycardia and vasodilatation. Psychological effects are variable but mostly consist of feelings of relaxation and well being. Perceptual disturbances are uncommon.

There is little evidence of tolerance in habitual users, in fact it appears that novices experience less effect and need to be sensitized. It is possible that this effect of reverse tolerance is due to the fact that the drug is accumulated avidly by fat (with a concentration up to 10-fold higher than in any other tissue) and is only slowly released from these stores. A regular user would thus more rapidly and easily achieve an effective concentration on future exposure.

Controversy surrounds the long term effects. There is little evidence for any causal relationship with chronic psychiatric illness, and the 'amotivational syndrome' reported is not supported by firm evidence.

9.8   Pharmacological effects

The major active constituent of cannabis is L-Δ^9-tetrahydrocannibinol (Δ^9-THC; Fig. 9.3) though Δ^{10}-THC is also active.

There are indications that Δ^9-THC interacts with biogenic amine systems in the brain, but the data are sufficiently conflicting as to make it impossible to form any

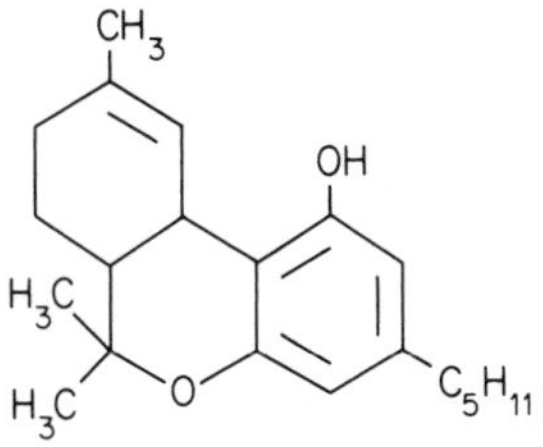

Fig. 9.3 Structure of Δ^9-tetrahydrocannabinol(Δ^9-THC)

satisfactory hypothesis. For example brain noradrenaline content has been variously reported to increase, decrease, or remain unchanged after Δ^9-THC. Brain 5-HT concentration has been reported to be increased and synthesis rate to decrease, but this has also been disputed. Uptake of both these transmitters is inhibited and the release facilitated in nerve endings (Johnson, Ho and Dewey, 1976).

THC also produces a decrease in the concentration of dopamine and acetylcholine in the brain.

One problem with many of these studies is the high doses of Δ^9-THC used. It is unlikely that the effects of a dose of Δ^9-THC of, for example, 20 mg/kg have much relevance to the mood changing effects of a relatively brief exposure to cannabis.

Reviews on the pharmacology of cannabis include those of Paton (1975) and Drew and Miller (1974).

PART 3   OPIATES

9.9   Historical aspects

Opium is collected from the seed heads of some of the poppy family especially *Papaver somniferum*. It has been used for millennia as a mood altering drug, and for centuries in a variety of medical remedies. The abuse of opiates has been considered a social problem for only a relatively short period of time, and legal controls introduced only during this century.

The active alkaloid of opium was isolated in 1803 in Germany by Sertürner who named it morphine (Fig. 9.4) after the Greek god of sleep, Morpheus. Heroin, diacetylmorphine, is a semisynthetic derivative first produced from morphine at

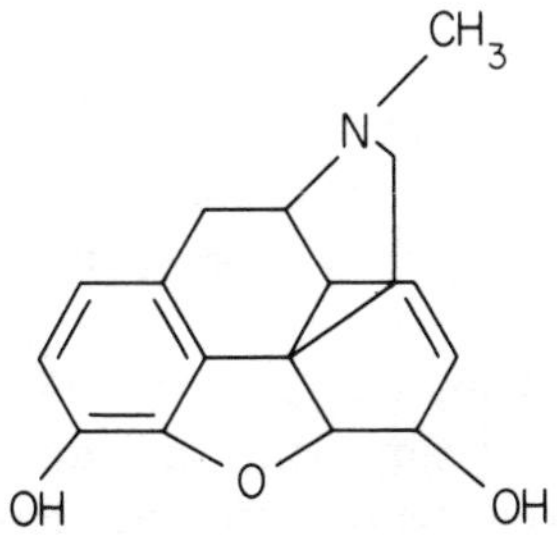

Fig. 9.4   Structure of morphine

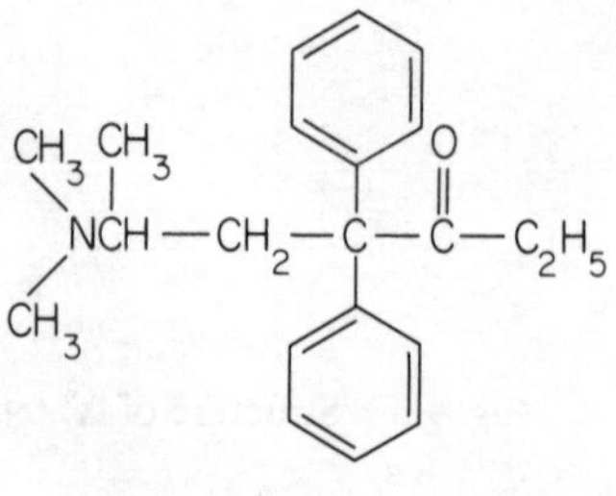

Fig. 9.5   Structure of methadone

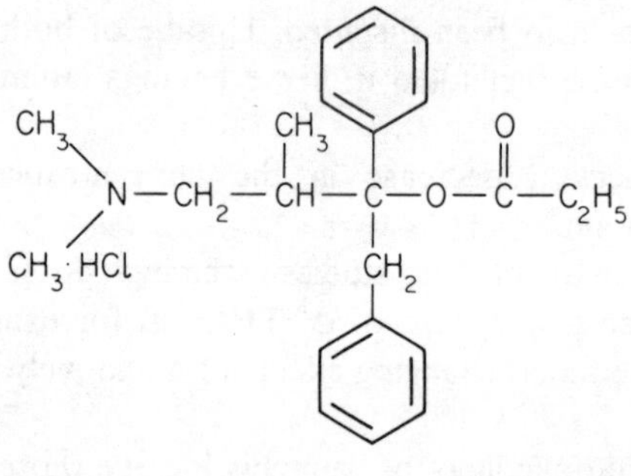

Fig. 9.6   Structure of dextropropoxyphene

St. Mary's Hospital London in 1874. It is ironic that it was introduced in 1898 as a remedy for morphine addiction. It was several years before it was realized that the reason it alleviated morphine addition was by substituting for it. To some extent history repeated itself in the 1950s when methadone was introduced as a treatment for addiction and found to have some serious addictive properties itself. It is not obviously structurally related to morphine (Fig. 9.5).

Various other opiate derivatives are in medical use. Codeine (methylmorphine) is a much weaker analgesic and does not easily produce dependence. Dextropropoxyphene is structurally similar to methadone (Fig. 9.6), is a useful analgesic and does not generally cause dependence.

Slang names for heroin include H, brown sugar, horse, and scag.

9.10   Physical and psychological effects

The prevalence of opiate addiction is very difficult to assess. Legal heroin production is small, but the international illegal traffic is extensive and lucrative, mostly deriving from regions in and around Turkey and China. It is estimated that there may be 100,000 addicts in the USA but many fewer, probably around 5,000 in the UK.

Contrary to the popular image of the 'junkie', studies in the UK have found that addicts are as likely to be in stable employment, and law abiding, as otherwise. Furthermore although there is undoubtedly a very high morbidity and mortality amongst users this is not invariably the case, and studies have found that one-third or more of addicts are abstinent two years later. Social factors are important, here, and in the special circumstances of Vietnam, military addicts almost all became abstinent on leaving the war.

Opiates may be taken by mouth (dropped), smoked or burned (inhaling the dragon), injected (fixed), intravenously (mainline), or subcutaneously (skin pop). Personal accounts of the effects are legion. There is an initial intense feeling of pleasure and well being, often compared to an orgasm. Other effects are analgesia, respiratory depression, hypotension, sleepiness, and dilated pupils. There may be little interest in social contact, and appetites for food and sex are diminished.

Tolerance to the effects of the drugs develops rapidly and a point is reached where the pleasurable effects are markedly diminished and use is maintained largely to avoid the withdrawal syndrome.

About 10 hours after the last dose the addict develops rhinorrhoea, lacrimation, yawning, sweating, and craving for the drug. He becomes restless and irritable, with nausea, vomiting, diarrhoea, cramps, and piloerection, explaining the slang 'cold turkey'. Hypertension, increased heart rate, and flushing also occur. After 72 hours the symptoms recede, but may take several further days to disappear.

This syndrome may also be brought about by use of a narcotic antagonist such as naloxone or levallorphan. However, following overdoses these drugs are invaluable for abolishing the respiratory depression, nausea, and other effects of morphine. In contrast naloxone has little effect in subjects who have not taken opiates making it a safe drug in patients in whom opiates are suspected but not confirmed. The short half-life of naloxone means that it must normally be given several times in the case of overdose.

The recent identification of specific opiate binding sites in the brain together with data on the endogenous ligand has provided a sound pharmacological explanation of the addictive properties of these drugs (see Sections 2.27–2.33).

9.11 Opiate dependence and withdrawal

Kosterlitz and Hughes (1975) suggested a hypothetical but entirely plausible physiological mechanism to explain the phenomenon of dependence and the withdrawal symptoms seen in opiate addicts.

Met-enkephalin (Sections 2.28–2.33) administration produces dependence in animals in exactly the same way as opiates and shows cross-tolerance with them. Evidence suggests that enkephalins are normally inhibitory to the release of certain other neurotransmitters.

When opiates are taken they will replace enkephalins at the sites of action and presumably generally at a higher concentration. This will produce tachyphylaxis or tolerance at the receptor site. Dependence will result since the neuronal systems will adapt as far as possible to deal with the opiates and withdrawal will perturb the systems.

The exogenously administered opiates inhibit the release of endogenous enkephalins. This has been shown by Simantov and Snyder (1976) who found that morphine dependent rats had elevated brain concentrations of enkephalin. Since cessation of firing of dopamine and acetylcholine neurons has been shown to be associated with increased concentrations of these transmitters, it seems reasonable to suppose that this phenomenon will exist for enkephalins and the elevated enkephalin concentration in morphine dependent rats indicates a shut-down of enkephalin function.

When, therefore, opiates are withdrawn, there is now no natural enkephalin inhibitory neurotransmitter system operating, it having been 'replaced' by the morphine. The result will be a marked 'overactivity' of a variety of other neurotransmitter systems, normally under enkephalinergic control. The result is the many physiological and psychological changes seen during withdrawal. This has been discussed in some detail by Snyder (1977).

PART 4   HALLUCINOGENS

9.12   Historical and social aspects

Hallucinogenic compounds have a long history of use in the form of plant derivatives. The peyotl cactus (mescaline) was known to Mexican Indians as was the mushroom *Psilocybe mexicana* (psilocybin and psilocin). In South America snuff from the cohida was used (dimethyltryptamine). The hallucinatory compound most used now is the synthetic lysergic acid diethylamide (LSD-25, LSD). Slang names include acid and sugar cubes.

LSD was discovered in 1943 during studies at Sandoz Laboratories in Switzerland on ergot alkaloids, which are of great pharmacological and historical interest. The pharmacological interest in these compounds results from their actions in contracting smooth muscle, particularly in blood vessels and the uterus (they are used to treat migraine and to initiate labour), their adrenergic blocking effects, and their central nervous system effects.

Historically reports on the effects of ergots go back to the Middle Ages in Europe when whole villages sometimes went 'insane' and were subsequently afflicted by St. Anthony's fire. This is the condition where the body extremities blacken as if burned and the whole phenomenon was due to a fungal contamination of rye grain that had been stored and subsequently baked in bread. A single village baker therefore would be the inadvertent cause of the disaster. The fungus was *Claviceps purpurea*, or ergot, extracts of which contain ergotamine, ergotoxine, and ergonovine. Ergotoxine is a mixture of ergocristine, ergocryptine, and ergocornine; structurally these compounds are related to lysergic acid and also to 5-HT as can be seen in Fig. 3.1.

The blackening of the extremities was gangrene resulting from vasoconstriction, and the madness the phenomena described below.

9.13   Physical and psychological effects of LSD

The physical effects of LSD usually appear 30–60 minutes after oral ingestion, peak after 3 hours and disappear in a further 3 hours. They include tachycardia, and a mild degree of hypertension, flushing, salivation, and occasionally motor disturbances.

The psychological changes last for rather longer but are usually over after 24 hours. The experience is one of heightened perception of sensory stimuli, with a breakdown of the boundary between the senses such that, for example, colours may be sensed as having a particular smell.

Perception of time may be distorted, usually as being lengthened, and similarly perception of space changed with objects appearing smaller or larger than in reality. Changes in mood are common but variable – the 'good' or 'bad' 'trip'.

Although the user usually retains insight with these experiences, occasionally he may acquire a false belief, such as having the ability to fly, with unfortunate consequences.

There is no clear withdrawal syndrome, perhaps reflecting the sporadic use of LSD, with little chance of tolerance developing.

The long term clinical effects are two-fold. First, there is the 'flashback phenomenon' where aspects of the 'trip' are re-experienced up to one year later without any further drug use. Second, there seems to be a small risk of a more prolonged schizophrenia-like reaction which may last for several years and require medical treatment, or may be self-limiting over a matter of weeks.

There is some evidence that LSD causes chromosomal damage, but whilst this has been demonstrated in laboratory animals, it is uncertain whether it occurs in humans.

9.14 Pharmacological effects of LSD

The LSD dose necessary for a trip is in the region of 50–200 µg and the drug is rapidly metabolized. This suggests that the profound mental changes are being produced by very few molecules of active compound.

Extensive studies have been made into the mechanism by which LSD produces its psychotic and hallucinogenic effects. The similarity of part of the structure to 5-HT and the fact that other methylated indoleamines produce hallucinations has resulted in much attention being focused on the effect of LSD on 5-HT systems. Indeed the fact that LSD could antagonize the action of 5-HT was reported by Gaddum in 1953.

Other data, however, have shown that LSD can act as a 5-HT agonist. Applied iontophoretically, LSD produces a similar inhibition in the firing of raphe neurons to that seen on application of 5-HT. Furthermore like other known 5-HT agonists administration of LSD leads to a slowing of the rate of 5-HT synthesis. LSD also acts as an agonist at postsynaptic receptors; however, so does 2-bromo-LSD, a non-hallucinogenic derivative. Some evidence suggests that LSD is far more potent on presynaptic 5-HT receptors (Aghajanian and Haigler, 1975) and that 2-bromo-LSD is ineffective presynaptically. LSD would therefore lead to a long lasting inhibition of 5-HT firing, and thus perhaps result in a disinhibition of other neuronal systems.

Recent studies, using both ligand binding techniques and behavioural models, have demonstrated that LSD acts on dopaminergic systems (Kelly and Iversen, 1975). Since both indoleamines and phenylethylamines can be hallucinogenic it is still unclear what mechanisms are involved in the hallucinogenic activity of LSD. Much effort has been expended in examining the structure and potency of hallucinogenic drugs but as yet no convincing theory on structure–activity relationships has emerged.

9.15 Other hallucinogenic compounds

As discussed in the general introduction several naturally occurring plants contain hallucinogenic compounds. Their physical and psychological effects are very similar to LSD and it is probable that most of them act pharmacologically in the same way since several of the active compounds are structurally related.

The first main group of compounds are the substituted indoles. This includes dimethyltryptamine, 5-methoxy,-N,N-dimethyltryptamine, and psilocin or 5-hydroxy-N,N-dimethyltryptamine. Structurally these compounds have much in common with LSD and, like LSD, will decrease the synthesis rate of 5-HT, suggesting that they also stimulate 5-HT receptors. Indeed these compounds produce behavioural changes in rats very similar to those seen when brain 5-HT concentrations are increased. One major difference between these drugs and LSD is that they act much more rapidly than ergot, but this may well be a pharmacokinetic phenomenon as LSD enters the brain less rapidly than the other compounds.

The other group of drugs abused are the substituted phenylethylamines, the best known drug being mescaline. Whether this drug acts to produce hallucination by actions on dopaminergic systems or because of a weak action on 5-HT neurotransmission (it is not a clinically potent drug) is uncertain. A related drug is 2,5-dimethoxy-4-methylamphetamine, referred to as DOM but also in street language as STP (a reference to the motor oil additive and also serenity, tranquillity, and peace). It can produce a LSD-like hallucination, but also several amphetamine-like physical changes.

Several of these drugs, for example psilocin, mescaline, and LSD, show cross-tolerance.

PART 5 PHENCYCLIDINE

9.16 Physical and psychological effects

In the Unites States this drug is now second only to marijuana as the most widely abused illicit drug.

The drug, often referred to as PCP (Fig. 9.7), was developed in the late 1950s and received attention as a possible anaesthetic. Administration to animals produces initial stimulation with higher doses producing sedation, apparent tranquillity, and then anaesthesia. It was rapidly removed from clinical use when it

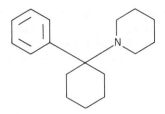

Fig. 9.7 Structure of phencyclidine

was found to produce hallucinations and confusion, but is still available commercially as a veterinary anaesthetic.

PCP is taken either on its own or combined with any of a variety of other illicit drugs. When smoked with marijuana it is known as superweed. On its own it may be smoked, eaten, sniffed, or injected and is known variously as angel dust, hog, goon, or T.

Small doses in humans produces a state of drunkenness. In higher but sub-anaesthetic doses it may produce a dissociative state where the subject is apparently conscious but is unresponsive to stimuli. Unlike other anaesthetics it stimulates both the respiratory and cardiovascular systems, even at anaesthetic doses.

The most serious effects, however, are a marked aggressiveness which may lead to murder or suicide, the most common cause of death following overdose. An acute schizophrenic-like state with hallucinations may be induced, and in some cases may last for some weeks. There is great variability in the elimination of PCP. Although its half-life is approximately 2 hours it seems that this is effectively extended in larger doses by gastroenteric recycling. PCP has been detected in urine up to 7 days after ingestion.

Dependence is said to occur, but it seems that little tolerance develops.

9.17 Pharmacological effects

In contrast to LSD and many other hallucinogens, PCP has no indole structure (Fig. 9.7). It is probable therefore that this drug has a mechanism of action very different from these other hallucinogens. This view is supported by the fact that no antagonists have been reported. It is a drug that effects the ion channels at nicotinic cholinergic receptors, but there have also recently been reports of specific PCP binding sites in the brain. However, the complexities of ligand binding studies means that such data must be treated with caution at this stage. Most of this current work has been reviewed by Snyder (1980).

PART 6   AMPHETAMINES

9.18 Physical and psychological effects

Amphetamines were developed from ephedrine, an alkaloid extracted from the plant *Ephedra vulgaris*. Several amphetamines have since been produced, including the dextro-isomer dexamphetamine, and methamphetamine.

Amphetamines were widely prescribed for depression, and general fatigue, in the 1940s and 1950s. After some years it became apparent that many users were psychologically dependent on the drugs, and also that the drugs were being widely used illegally. Voluntary restrictions by the medical profession resulted in the virtual abolition of amphetamine prescribing, but the drugs are easily synthesized and remain much used illegally.

Slang terms are derived from approved or trade names, 'dexies' (dexamphetamines) and 'bennies' (Benzedrine), from their actions, 'speed', 'pep pills', or 'uppers', or from their appearance, 'oranges'.

Oral ingestion is followed by feelings of well being, excitement, and increased concentration, energy, and activity. Increasing doses lead to tachycardia, dilated pupils, and increased reflexes, later to tremor, ataxia, nausea, and headache, and in extreme cases cerebral haemorrhage and neurological damage.

Although tolerance does not seem to develop there is often a rapid increase in dose taken, or a change from oral to intravenous administration. A bout of drug taking may last several days and be terminated only by extreme fatigue. On withdrawal users are hungry, lethargic, and often very depressed, and it has been argued that this represents a withdrawal syndrome rather than a physiological swing.

It is well recognized that chronic high dose use leads to the 'amphetamine psychosis'. This is a state which, when fully developed, is indistinguishable from a schizophrenic illness (see Section 5.15), but which usually resolves over days as the drug is excreted. Some individuals, however, fail to recover and are left with a residual illness. It is controversial whether this state is caused by amphetamines, or is merely precipitated by them in susceptible persons.

9.19 Pharmacological aspects

The neuropharmacology of amphetamines (see Fig. 5.4) has been extensively studied in experimental animals.

At low doses in rats the drugs produce arousal and locomotor activity, the activity being very characteristic with the back arched and the head down, together with bursts of rearing and sniffing. As the dose of amphetamine increases the locomotor activity decreases being replaced by stereotyped behaviour or 'stereotypy'. The behaviour includes sniffing, chewing, licking, and biting and it is presumably the increase of these behaviours which causes a suppression of locomotion rather than that locomotion is not produced by high doses of the drugs. In some elegant lesioning experiments it was shown that the locomotion is due predominantly to dopamine release in the mesolimbic forebrain (n. accumbens) whilst the stereotypy is due to release of dopamine in the caudate (Kelly, Seviour and Iversen, 1975).

In low doses the major effects of amphetamine is to release dopamine. It does not produce behavioural responses in rats depleted of catecholamines by administration of the synthesis inhibitor α-methyl-*p*-tyrosine or following destruction of dopaminergic terminals by injection of the neurotoxin 6-hydroxydopamine (6-OHDA). Reserpine pretreatment has much less effect on amphetamine-mediated behaviour; however, if we assume that amphetamine produces a preferential release of the newly synthesized monoamine then this result becomes easier to explain since reserpine does not effect these pools. The behaviours are also inhibited by the neuroleptics, including some fairly specific dopamine antagonists.

In addition amphetamine can inhibit reuptake of monoamines by the nerve ending, which again would have the effect of potentiating dopamine-mediated behaviour.

At very high dose amphetamines will release 5-HT and noradrenaline as well as dopamine. One specific amphetamine, *p*-chloroamphetamine, has the predominant effect of releasing 5-HT and the behavioural effects which follow this drug are those seen after administration of 5-HT agonists. In common with several other amphetamines this drug is also a weak MAO inhibitor.

PART 7   COCAINE

9.20   Physical, psychological, and pharmacological effects

Cocaine, popularly known as 'coke', 'charlie', 'big C', or 'snow' is obtained from the plant *Erythroxylum coca*. The stimulant effects of the drug have been known for centuries in South America where the leaves were chewed to increase physical endurance.

Pure cocaine was isolated in 1858 and within a few years was being used clinically as a local anaesthetic. This use has been largely superseded by derivative drugs, but it is also used as an adjunct to opiate treatment for chronic pain, although its value here has been questioned (Twycross, 1977).

Illegal use as a stimulant drug has been well documented, and it has been popular throughout the twentieth century. Partly because of the lore surrounding its use, and its current expense, it is considered a 'high class' drug.

Cocaine is usually sniffed, and absorbed quickly, through the nasal mucosa. It promotes a feeling of well being and euphoria, with heightened awareness and increased energy. Cocaine is rapidly metabolized and the effects last usually for less than one hour.

There is little evidence for dependency, and none for the development of tolerance, indeed there is in some people an apparent increased sensitivity. Larger doses may produce a paranoid psychosis, often with tactile hallucinations (cocaine bug), but this is usually self-limiting. Chronic use may lead to ischaemia and ulceration of the nasal mucosa through the local vasoconstrictive action of cocaine.

Pharmacologically cocaine has been shown to inhibit neuronal uptake of noradrenaline and dopamine. It is not however an antidepressant (see Section 3.25).

PART 8   BENZODIAZEPINES

9.21   Problems of dependence

In Chapter 4 (Sections 4.5–4.11) the pharmacological aspects of the benzodiazepines were examined. In this section only the problems of dependence will be examined.

Benzodiazepines are the world's most widely prescribed group of drugs and it is estimated that over 40 billion doses are consumed daily. Partly because of this and also because of their limited psychotropic effect, illicit use is rare.

Although there are minor differences between different compounds, their metabolism is closely inter-related and the main difference is that of duration of action (Section 4.9). The action of the benzodiazepines is to reduce the subjective feeling of anxiety, and to induce sedation.

It is the prevalence of the stress of life, now seen to merit drug treatment, and the high degree of safety of the drugs that contribute to their widespread use. Although they have been used for over 20 years it is only recently that the problem of dependence has been widely recognized, although it was first reported by Hollister, Motzenbecker and Degan in 1961. The use of benzodiazepines should ideally be limited to short term treatment of anxiety or insomnia. However, many patients are on the drugs for years, and so at risk of becoming dependent. The risk has been estimated at one per 5×10^6 patient months, which is small, but represents many thousand cases per year worldwide.

It is difficult to assess the development of tolerance, as anxiety and thus the need for drugs fluctuates, but it seems likely that it does occur to some extent, and that some patients require increasing doses. Studies on withdrawal of the drug following 20 weeks of treatment, suggest that patients experience trembling, poor appetite, and faintness and if taking high doses, twitching, shaking, palpitations, hypersensitivity to noise and touch, and anxiety and insomnia. These symptoms are relieved by taking benzodiazepines, suggesting that they represent a true withdrawal syndrome.

PART 9   BARBITURATES

9.22   Physical and psychological effects

Barbiturates were introduced in 1903 as sedative drugs. They were widely prescribed as a treatment for insomnia, depression, anxiety, and short acting derivatives were used as anaesthetics.

In the 1950s and 1960s barbiturate dependency was a major health problem in the UK with over half a million regular takers, 20% of whom were probably dependent. Concern over this, together with the belief that the new benzodiazepines were effective and safe, resulted in a major change in prescribing habits. Barbiturates are now rarely prescribed as sedatives, but supplies are still available, and illicit use continues.

The main action of barbiturates is central nervous system depression, with initially slight impairment of function but higher doses producing 'drunkenness' and unconsciousness. They reduce the rapid eye movement (REM) stage of sleep, with reduction in dreaming. Tolerance develops readily, and withdrawal symptoms are seen when the drug is stopped. Initially tremor, agitation and perceptual disturbances occur, with more severely delirium and convulsions. These convulsions may occur some time after stopping the drug, reflecting the long duration of action of some of the barbiturates. REM sleep is increased, and increased amount and vividness of dreaming is a disturbing symptom.

Slang names include barbs, downers, goofballs, phennies, sleeping pills, and hearts.

9.23 Pharmacological aspects

The barbiturates are all derivatives of barbituric acid, which has no sedative action. The first compound used clinically was barbitone (or barbital: substitute the -one with -al for naming in the USA in the case of all these drugs), followed soon after by phenobarbitone. Other shorter action drugs subsequently became available including pentobarbitone. The duration of action of the barbiturates is inversely related to their lipid solubility. The structure of some of the barbiturates is shown in Fig. 9.8.

The short acting barbiturates such as methohexitone and thiopentone are still used as intravenous anaesthetics. The longer acting drugs are sometimes used as sedative–hypnotics. Phenobarbitone is still occasionally used in the treatment of generalized tonic/clonic seizure disorders. Its anticonvulsant properties are not due to a sedative action, since it is effective at non-sedative doses.

The mechanism of action is still unclear. Pentobarbitone inhibits the uptake of calcium into the nerve ending (Blaustein and Ector, 1975) and calcium is intimately connected with the release of neurotransmitters (see Rubin, 1970). The drugs also have local and general anaesthetic actions (see Richards, 1972).

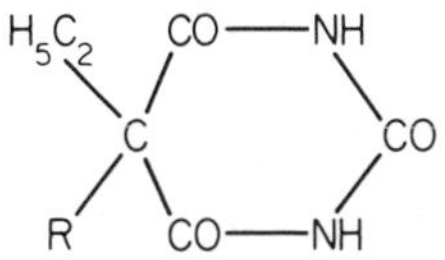

Name	R
Barbitone	C_2H_5
Phenobarbitone	C_6H_5
Pentobarbitone	1-Methylbutyl
Amylobarbitone	Isoamyl

Fig. 9.8 Structures of some common barbiturates

9.24 Metabolism

In general, the drugs are absorbed rapidly. Metabolism can follow four paths:

1. oxidation of radicals in the 5 position;
2. removal of N-alkyl groups;
3. ring cleavage; and
4. conversion of thiobarbiturates to the oxygen analogues.

Barbiturate metabolism is very dependent on the microsomal drug metabolizing enzyme systems. Thus the rate of inactivation can be altered by administration of other drugs or by hepatic failure and possible drug interactions have to be carefully considered and evaluated before use.

PART 10   NICOTINE (TOBACCO)

9.25   Physical and psychological effects

Nicotine is the major pharmacologically active component of tobacco. When tobacco was first introduced into Europe in the sixteenth century its use was regarded as antisocial, and it was not until the First World War that its use became widespread, with the advent of cigarette smoking. It is only recently that smoking has again come to be regarded as an antisocial habit and this is perhaps associated with the recognition that it is a major cause of mortality, through bronchitis, lung cancer, and cardiovascular disorders. There is also evidence that exposure to tobacco smoke causes health problems in non-smokers. Furthermore it is now realized that smoking produces addiction.

Nicotine is taken mostly by inhalation of burning leaves, but also through chewing. Absorption is rapid from the bronchial mucosa and lung, taking only seconds to reach the brain. It raises cerebral arousal and stimulates the cardiovascular system, both directly and through the release of noradrenaline. It also relaxes muscle, and is thus rare in increasing arousal whilst also increasing relaxation. This is clearly very powerful in promoting psychic dependence, but physical dependence is also seen.

Tolerance certainly occurs, as in the novice, nicotine intake causes nausea, palpitations, dizziness, and sweating, in doses which would scarcely affect the habitual user.

There is also a clear withdrawal syndrome of craving, depression, tension, restlessness, lowered heart rate and blood pressure, and sleep disturbances. In spite of these clear charges it is likely that the social and psychological aspects of dependence are the more powerful factors in making smoking difficult to give up.

9.26   Mechanism of dependence

Nicotine is an extremely toxic drug. Most of the peripheral effects it produces (on the cardiovascular system etc.) are precisely those seen during smoking.

Nicotine, of course, is an agonist at acetylcholine receptors of the nicotinic type (Section 2.18) and its effects are those of cholinergic stimulation. Continued use presumably produces tolerance at the receptor site and inhibits the release of endogenous transmitter. In the absence of nicotine there is insufficient neurotransmitter and physiological withdrawal phenomena occur. This phenomenon is similar to that postulated for opiate withdrawal (Section 9.11).

Common Abbreviations Used in Neuropharmacology

A = adrenaline (epinephrine)
acetyl CoA = acetyl coenzyme A
ACh = acetylcholine
AChE = acetylcholinesterase
ACTH = adrenocorticotropic hormone
$ACTH_{4-10}$ = fragment 4–10 of
    ACTH
ADP = adenosine diphosphate
ADTN = 2-amino-6,7-dihydroxy-
    1,2,3,4-tetrahydronaphthalene
AMP = adenosine monophosphate
AMPT = α-methyl-*p*-tyrosine
AOAA = amino-oxyacetic acid
APUD = amine precursor uptake
    and decarboxylation
ATP = adenosine-5′-triphosphate
ATPase = adenosine triphosphatase

BOL = 2-brom-LSD

Ca^{2+} = calcium ion
cAMP = cyclic 3′,5′-adenosine
    monophosphate
CAT = choline acetyltransferase
CDP = chlordiazepoxide
cGMP = cyclic GMP = guanosine-
    3′,5′-monophosphate
ChE = cholinesterase
CNS = central nervous system
COMT = catechol-*O*-methyltransferase
CPZ = chlorpromazine
CRF = corticotropin-releasing
    factor
CSF = cerebrospinal fluid

cyclic AMP = cyclic 3′,5′-adenosine
    monophosphate

DA = dopamine
db-cAMP = dibutyrylcyclic AMP
DBH = dopamine-β-hydroxylase
5,6-DHT = 5,6-dihydroxytryptamine
5,7-DHT = 5,7-dihydroxytryptamine
$DMPH_4$ = 6,7-dimethyl-5,6,7,8-
    tetrahydropterin
DOMA = 3,4-dihydroxymandelic acid
DOPA = dihydroxyphenylalanine
DOPAC = dihydroxyphenylacetic acid
DOPS = dihydroxyphenylserine

E = epinephrine (adrenaline)
ECS = electroconvulsive shock
ECT = electroconvulsive therapy
EDTA = ethylene diamine tetra-
    acetic acid
EEG = electroencephalogram
EGTA = ethylene glycol *bis*-(β-
    aminoethyl ether) tetraacetic acid
EPSP = excitatory postsynaptic
    potentials

GABA = γ-aminobutyric acid
GAD = L-glutamate decarboxylase
    = glutamic acid decarboxylase
GLC = gas chromatography = gas/
    liquid chromatography
GC/MS = gas chromatography/mass
    spectroscopy
GH = growth hormone

Glu = glutamate
GTP = guanosine triphosphate

5-HIAA = 5-hydroxyindoleacetic
　　acid
HIOMT =hydroxyindole-O-methyl-
　　transferase
5-HT = 5-hydroxytryptamine =
　　serotonin
5-HTOH = 5-hydroxytryptophol
5-HTP = 5-hydroxytryptophan
HVA = homovanillic acid

IAA = indoleacetic acid
IC_{50} = concentration for 50%
　　inhibition
IPSP = inhibitory postsynaptic
　　potentials

K^+ = potassium ion
K_a = association constant
K_d = dissociation constant
K_i = inhibition constant
K_m = Michaelis constant

LC = locus coeruleus
Leu = leucine
LH = luteinizing hormone
LHRH = luteinizing hormone
　　release hormone
Li^+ = lithium ion
LPH = β-lipotropin
LSD = LSD_{25} = D-lysergic acid
　　diethylamide

MAO = monoamine oxidase
MAOI = monoamine oxidase inhibitor
5-MeO-DMT = 5-methoxy-N,N-
　　dimethyltryptamine
Met = methionine
Mg^{2+} = magnesium ion
MHPG = 3-methoxy-4-hydroxy-
　　phenylglycol, also referred to
　　as MOPEG
MOPEG = 3-methoxy-4-hydroxy-
phenylglycol, also referred to
　　as MHPG
MTA = 3-methoxytyramine

NA =noradrenaline (norepinephrine)
Na^+ = sodium ion
NAD = nicotinamide-adenine
　　dinucleotide
NE = norepinephrine (noradrenaline)
NEFA = non-esterified fatty acids
NM = normetanephrine

6-OHDA = 6-hydroxydopamine

PCPA = p-chlorophenylalanine
PEA = phenethylamine
PGE_1 = prostaglandin E_1
Phe = phenylalanine
PRL = prolactin
PTZ = pentylenetetrazol

REM = rapid eye movement (sleep)

SAM = S-adenosylmethionine
s.d. = standard deviation
s.e.m. = standard error of the mean
SP = substance P
SRIF = somatotropin release-
　　inhibiting factor

$T_{\frac{1}{2}}$ = half-life
T_3 = L-tri-iodothyronine
T_4 = thyroxine
TBZ = tetrabenazine
TCA = tricarboxylic acid
Δ^9-THC = Δ^9-tetrahydrocannabinol
TLC = thin-layer chromatography
TRH = thyrotropin releasing
　　hormone
Trp =Tryptophan
TSH = thyroid stimulating
　　hormone = thyrotropin
Tyr = tyrosine

VMA = 3-methoxy-4-
　　hydroxymandelic acid
V_{max} = maximum velocity

Drugs Cited in Text with Some of their Common Trade Names

Drug	UK trade name	USA trade name
Amantadine	Symmetrel	Symmetrel
Amitriptyline	Tryptizol, Lentizol	Elavil, Endep
Amphetamine	Benzedrine	Benzedrine
Amylobarbitone	Amytal	Amytal
Baclofen	Lioresal	—
Benzhexol	Artane	Artane
Benztropine	Cogentin	Cogentin
Bromocriptine	Parlodel	—
Butriptyline	Evadyne	—
Butobarbitone	Soneryl	Butisol
Chlordiazepoxide	Librium	Librium, Libritabs
Chlorimipramine	Anafranil	—
Chlormethiazole	Heminevrin	—
Chlorpromazine	Largactil	Thorazine
Cimetidine	Tagamet	Tagamet
Clobazam	Frisium	—
Clonidine	Catapres, Dixarit	—
Clopenthixol	Clopixol	—
Cyproheptadine	Periactin	Periactin
Desimipramine	Pertofran	Pertofran, Norpramin
Dexamphetamine	Dexedrine	Desoxyn, Dexedrine
Diazepam	Valium	Valium
Dibenzepin	Noveril	—
Disulfiram	Antabuse	Antabuse
L-Dopa (Levodopa)	Larodopa	Larodopa
L-Dopa with peripheral decarboxylase inhibitor	Madopar (+benserazide) Sinemet (+carbidopa)	Madopar Sinemet
Dopamine	Intropin	Intropin
Doxepin	Sinequan	Sinequan
Droperidol	Droleptan	—
Ethopropazine	—	Parsidol

Drug	UK trade name	USA trade name
Flunitrazepam	—	Rohypnol
Flupenthixol	Depixol, Fluanxol	—
Fluphenazine	Modecate	Prolixin, Permital
Flurazepam	Dalmane	Dalmane
Haloperidol	Serenace, Haldol	Haldol
Hexobarbitone	Evidom	Evipal
Imipramine	Tofranil	Tofranil
Iprindole	Prondol	—
Iproniazid	Marsilid	Marsilid
Isocarboxazid	Marplan	Marplan
Isoprenaline	Suscardia	Isoprel
Lithium carbonate	Priadel, Phasal, Camcolit	Eskalith, Lithonate, Lithane
Lorazepam	Ativan	—
Maprotiline	Ludiomil	—
Medazepam	Nobrium	—
Meprobamate	Miltown, Equanil	Miltown, Equanil
Methadone	Physeptone	Dolophine
Methylphenidate	Ritalin	Ritalin
Methysergide	Deseril	Sansert
Metoclopramide	Maxolon	—
Mianserin	Norval, Bolvidon	—
Molindone	—	Lidone, Moban
Naloxone	Narcan	Narcan
Nitrazepam	Mogadon	Mogadon
Nomifensine	Merital	—
Nortriptyline	Aventyl	Aventyl
Oxazepam	Serenid	Serax
Pentobarbitone	Nembutal	Nembutal
Perphenazine	Fentazin	Trilafon
Phenelzine	Nardil	Nardil
Phenobarbitone	Luminal	Luminal
Phenoxybenzamine	Dibenyline	Dibenyline
Phenylephrine	Neophryn, Prefrin	Isophrin, Neosynephrine
Pimozide	Orap	—
Prochlorperazine	Stemetil	Compazine
Promazine	Sparine	Sparine
Promethazine	Phenergan, Avomine	Phenergan
Propranolol	Inderal	Inderal
Protriptyline	Concordin	Vivactil
Reserpine	Serpasil	Serpasil
Secobarbitone	Seconal	Seconal
Succinylcholine	Anectine	Anectine

Drug	UK trade name	USA trade name
Temazepam	Euhypnos, Normison	Cerepax, Levanxol
Tetrabenazine	Nitoman	Nitoman
Thiopentone	Pentothal	Pentothal
Thioridazine	Mellaril	Mellaril
Tranylcypromine	Parnate	Parnate
Triazolam	Halcion	—
Trifluoperazine	Stelazine	Stelazine
L-Tryptophan	Optimax (+pyridoxine)	—
	Pacitron	
Viloxazine	Vivalan	—

This list is not comprehensive, some drugs are sold under several trade names or are included in joint preparations. In this list the best known trade names have generally been cited. Not all the drugs listed are sold both in Europe and North America.

Drug Trade Names with their Non-Proprietary Names

Trade name	Non-proprietary name
Amytal	Amylobarbitone
Anectine	Succinylcholine
Anafranil	Chlorimipramine
Antabuse	Disulfiram
Artane	Benzhexol
Ativan	Lorazepam
Aventyl	Nortriptyline
Avomine	Promethazine
Benzedrine	Amphetamine
Bolvidon	Mianserin
Butisol	Butobarbitone
Camcolit	Lithium carbonate
Catapres	Clonidine
Cerepax	Temazepam
Clopixol	Clopenthixol
Cogentin	Benztropine
Compazine	Prochlorperazine
Concordin	Protriptyline
Dalmane	Flurazepam
Depixol	Flupenthixol
Deseril	Methysergide
Desoxyn	Dexamphetamine
Dexedrine	Dexamphetamine
Dibenyline	Phenoxybenzamine
Dixarit	Clonidine
Dolophine	Methadone
Droleptan	Droperidol
Elavil	Amitriptyline
Endep	Amitriptyline
Equanil	Meprobamate
Eskalith	Lithium carbonate

Trade name	Non-proprietary name
Euhypnos	Temazepam
Evadyne	Butriptyline
Evidom	Hexobarbitone
Fentazin	Perphenazine
Fluanxol	Flupenthixol
Frisium	Clobazam
Halcion	Triazolam
Haldol	Haloperidol
Heminevrin	Chlormethiazole
Inderal	Propranolol
Intropin	Dopamine
Isophrin	Phenylephrine
Isuprel	Isoprenaline
Largactil	Chlorpromazine
Larodopa	L-Dopa (Levodopa)
Lentizol	Amitriptyline
Levanxol	Temazepam
Libritabs	Lithium carbonate
Librium	Chlordiazepoxide
Lidone	Molindone
Lioresal	Baclofen
Lithane	Lithium carbonate
Lithonate	Lithium carbonate
Ludiomil	Maprotiline
Luminal	Phenobarbitone
Madopar	L-Dopa (+benseraside)
Marplan	Isocarboxazid
Marsilid	Iproniazid
Maxolon	Metoclopramide
Mellaril	Thioridazine
Merital	Nomifensine
Miltown	Meprobamate
Moban	Molindone
Modecate	Fluphenazine
Mogadon	Nitrazepam
Narcan	Naloxone
Nardil	Phenelzine
Nembutal	Pentobarbitone
Neophryn	Phenylephrine
Neosynephrine	Phenylephrine
Nitoman	Tetrabenazine
Nobrium	Medazepam
Normison	Temazepam
Norpramin	Desimipramine
Norval	Mianserin

Trade name	Non-proprietary name
Noveril	Dibenzepin
Optimax	L-Tryptophan (+pyridoxine)
Orap	Pimozide
Pacitron	L-Tryptophan
Parlodel	Bromocriptine
Parnate	Tranylcypromine
Parsidol	Ethopropazine
Pentothal	Thiopentone
Periactin	Cyproheptadine
Permitil	Fluphenazine
Pertofran	Desimpramine
Phasal	Lithium carbonate
Phenergan	Promethazine
Physeptone	Methadone
Prefrin	Phenylephrine
Priadel	Lithium carbonate
Prolixin	Fluphenazine
Prondol	Iprindole
Ritalin	Methylphenidate
Rohypnol	Flunitrazepam
Sansert	Methysergide
Seconal	Secobarbitone
Serax	Oxazepam
Serenace	Haloperidol
Serenid	Oxazepam
Serpasil	Reserpine
Sinemet	L-Dopa (+carbidopa)
Sinequan	Doxepin
Soneryl	Butobarbitone
Sparine	Promazine
Stelazine	Trifluoperazine
Stemetil	Prochlorperazine
Suscardia	Isoprenaline
Symmetrel	Amantadine
Tagamet	Cimetidine
Thorazine	Chlorpromazine
Tofranil	Imipramine
Trilafon	Perphenazine
Tryptizol	Amitriptyline
Valium	Diazepam
Vivactil	Protriptyline
Vivalan	Viloxazine

This list is not comprehensive. Some drugs are sold under several trade names. In this list the best known names have generally been listed.

Appendix 4

Neurotransmitters; Metabolic Inhibitors, Agonists, and Antagonists

Transmitter	Synthesis inhibitors	Degradation inhibitors	Receptor type	Agonists	Antagonists	Notes
5-Hydroxy-tryptamine	*p*-Chlorophenylala-nine, 6-fluorotrypto-phan (Trp hydroxy-lase inhibitors)	Tranylcypromine, pargyline, phenelzine, iproniazid (monoamine oxidase inhibitors)	5-HT	5-methoxy-*N,N*-dimethyltryptamine, quipazine, 5-hydroxytryptamine	Methysergide, methergoline, cyproheptadine, cinanserin, mianserin	LSD can be an agonist or antagonist. L-Tryptophan loading increases 5-HT formation. Two 5-HT receptor types may exist (Peroutka and Snyder, 1979)
Dopamine	α-Methyl-*p*-tyrosine (tyrosine hydroxylase inhibitor)	Monoamine oxidase inhibitors listed above	Non-selective	Apomorphine, ADTN, dopamine	Chlorpromazine, fluphenazine, and most neuroleptics	Amphetamine acts as an agonist in so far as it releases dopamine from the nerve endings. See also paper by Kebabian and Calne (1979) on D_1 and D_2 receptor types
			D_1	Not known	α-Flupenthixol (relatively specific)	
			D_2	Bromocriptine	Metoclopramide, sulpiride	

Transmitter	Synthesis inhibitor	Metabolism/degradation inhibitor	Receptor	Agonist	Antagonist	Comments
Noradrenaline (adrenaline)	Disulfiram FLA-63 (dopamine-β-hydroxylase inhibitors)	Monoamine oxidase inhibitors listed above (catechol-O-methyltransferase is inhibited by tropolone)	Non-selective α	Adrenaline, noradrenaline	Labetalol, dihydroergocryptine, phenoxybenzamine, phentolamine	Labetalol also inhibits α-receptors
			α₁	Methoxamine, phenylephrine	Prazosin, WB 4101	
			α₂	Clonidine	Yohimbine, piperoxan	
			Non-selective β	Isoprenaline (isoproterenol), adrenaline	Propranolol, alprenolol	
			β₁	Not known	Practolol, atenolol	
			β₂	Salbutamol	Butoxamine	
Histamine	α-Hydrazino-histidine (L-histidine decarboxylase inhibitor)	Isoniazid, hydralazine (diamine oxidase inhibitors)	Non-selective	Histamine	Promethazine	L-Histidine loading increases histamine formation
			H₁	2-Thiazolylethylamine	Mepyramine	
			H₂	Dimaprit	Metiamide, cimetidine	
Acetylcholine	Hemicholinium (high affinity choline uptake inhibitor)	Physostigmine di-isopropylphospho-fluoridate (acetylcholinesterase inhibitors)	Non-selective	Acetylcholine	Atropine, hexamethonium	
			Muscarinic	Muscarine, methacholine	d-Tubocurarine, quinuclidinyl benzoate Dihydro-β-erythroidine	
			Nicotinic	Nicotine		
GABA	Allylglycine (glutamic acid decarboxylase inhibitor)	Amino-oxyacetic acid (GABA transaminase inhibitor)	GABA	Muscimol, GABA	Bicuculline, picrotoxin, pentylenetetrazol	Probably only bicuculline acts at the receptor. Benzodiazepines increase GABA function
Enkephalin	Cycloheximide (protein synthesis inhibitor)	Not known	Enkephalin	Met-enkephalin, morphine	Naloxone	See article by Kosterlitz (1980) on different types of opiate receptor and their proposed agonists and antagonists
Glycine	Not known	Not known	Glycine	Glycine	Strychnine	

Glossary

For specific neuroanatomical terms and metabolic enzymes see also Chapter 2. For specific drug names and abbreviations see appendices 1–3. The glossary should be used in conjunction with the index.

Affect Mood, feelings, or emotions

Affinity Attraction between ligand and receptor, quantified in terms of the affinity constant

Agnosia Loss of the ability to recognize sensory stimuli

Agonist A compound which acts at a receptor to produce the effect of the natural ligand

Akinesia Absence (or reduction) of voluntary muscular movement

Alkaloid Organic basic compounds found in plants

Alkyl A prefix referring to a radical derived form an open chain hydrocarbon, often referred to as aliphatic

Antagonist A compound which acts at a receptor to antagonize the effect of the natural ligand

Antinociceptive Having the action of reducing or abolishing painful stimuli

APUD cells Amine precursor, uptake and decarboxylation cells, derived from the ectoblast, see Pearse (1969)

Arteriosclerosis The thickening, hardening, and loss of elasticity of arteries

Aryl A prefix referring to a radical derived from benzene, often referred to as 'aromatic'

Ataxia Loss of muscular coordination

Autopsy Post-mortem

Autoreceptor A receptor situated on the presynaptic nerve ending, sensitive to the transmitter released by the neuron. Also called presynaptic receptors.

Autosome Chromosome not determinant of sexual differentiation. Autosomal dominance: inheritance of non-sex-linked genetic factors where the gene needs to be present on only one chromosome

Basal ganglia A series of brain stem nuclei including corpus striatum (putamen and globus pallidus) and substantia nigra

Bipolar Affective illness where both depression and mania occur

Bound 1. Association of drugs with plasma proteins, also applying to tryptophan. 2. The combination of ligand with receptor 'specific bound'. 3. The combination of ligand with non-receptor material 'non-specific bound'

Carcinoid syndrome Medical condition with a 5-hydroxyindole secreting tumour, usually located in the gastrointestinal system

Catalepsy   A state of rigidity with either resistance to alteration or ready adoption of a new imposed posture. Often used to refer to animal behaviour

Catatonia   A clinical symptom which may be either a marked reduction or increase in mobility or an alternation between the two. May also describe automatism or stereotyped movements

Catechol   1,2-dihydrobenzene. Catecholamines, group of compounds containing the catechol structure; see also monoamines

Cerebrospinal fluid   Fluid bathing the brain and spinal cord and which is also contained in the ventricles. Sampled either by lumbar puncture or more rarely from the ventricles

Chelation   A chemical reaction in which a metallic ion is sequestered and therefore inactivated

Chorea   Repetitive involuntary jerky movements

Circling   Behaviour elicited by amphetamines or dopamine agonists in animals with a unilateral lesion of the nigrostriatal pathway

Clearance   A measure of the rate of elimination of drug from the body

Cofactor   Compound or ion, not itself involved in the reaction, but facilitating an enzyme reaction

Compartments   Areas of drug or neurotransmitter distribution having different kinetic characteristics

Competitive inhibition   Inhibition that is dependent on the concentration of inhibitor and substrate

Confabulation   Fabrication invented to fill gaps in memory

Corpus striatum   Area of the brain, part of the basal ganglia consisting of the caudate nucleus and putamen

Delusion   A belief held without any supportive evidence, usually but not necessarily false

Depersonalization   The subjective experience that one's body is unreal or changed in some way

Desensitization   1. An adaptation of the receptor to reduce the function of the system. 2. A psychological treatment designed to reduce anxiety to specific situations, by controlled exposure to them

Diencephalon   Anterior part of brain stem including hypothalamus, thalamus, and posterior pituitary

Dissociation constant   Measure of the tendency of compounds to separate. In ligand binding studies a measure of the attraction between ligand and receptor, the reciprocal of the affinity constant

Dizygotic   In twin studies, refer to twins who have developed from two ova and therefore have different genetic characteristics

Down-regulation   An adaptation usually of a receptor, to reduce the function of the system

Dyskinesia   Impairment of voluntary movements

Dysphasia   Impairment of language

Dyspraxia   Impairment of the ability to perform coordinated movement

Edema   See oedema

Electrolytic lesions   Destruction of specific neuronal pathways by passage of electricity between electrodes inserted into the brain

Enzyme induction   Increase in potential activity of an enzyme

Extrapyramidal   Refers to motor control not involving pyramidal tracts, having its origin in the basal ganglia

Flight of ideas   Rapid succession of thoughts without logical connections

Free   Form of compound not bound (q.v.)

Gas chromatography   Separation of volatile compounds by injection into a gas stream which is percolated over a stationary phase, which may be either a solid or a liquid (which is spread as a thin film over an inert solid). The separated compounds then pass into a suitable detector

Glial cells   Supporting cells within the central nervous system, thought not to be directly involved in neurotransmission

Globus pallidus   Specific nucleus within the basal ganglia

Grand mal   Major seizure disorder with tonic and clonic muscular movements and loss of consciousness

Hallucination   Sensory perception not based on a real stimulus

Half-life   Time taken for the concentration of a compound to decrease by 50%

Histochemistry   Study of composition of cells using chemical techniques. Fluorescence histochemistry involves reacting the monoamine neurotransmitters with chemicals to produce a fluorescent derivative that can be mapped.

Hyperbaric   Increased pressure – hyperbaric oxygen is raised oxygen pressure

Hyperkinesia   Increased movements or activity

Hypertension   Raised blood pressure

Hypnotic   Sleep inducing

Hypochondriasis   Over-concern about health

Hypophysis   Pituitary. Hypophysectomy is the removal of the pituitary

Hypotension   Lowered blood pressure

Hypothermia   Low body temperature

Ideas of reference   Ideas that normal events have relevance to or are commenting on oneself

Immunofluorescence   Fluorescence histochemistry using antibodies to identify the compounds under investigation

Indoles   2,3-Benzopyrrole. Indoleamines, group of compounds containing the indole structure; see also monoamines.

Infarct   An area of tissue death due to reduced blood supply

Ion channel   Channel concerned with the passage of ions across a cell membrane

Iontophoresis   Administration of compounds through micropipettes which are released by an electric current

Isomerism   1. Structural: the possession of 2 or more compounds of the same molecular formula but different structures. 2. stereo: two or more compounds possessing the same molecular and structural formulas but having different spatial configurations

Lacrimation   The production of tears, not necessarily associated with sadness

LD$_{50}$ Dose of drug which is lethal to 50% of the test population (Lethal Dose 50)

Libido   Sexual drive

Life events   Experiences which are part of normal life, but are stressful and thought to be involved in the precipitation of psychiatric disorders

Ligand   Compound which specifically binds to a receptor

Lumbar puncture   The sampling of cerebrospinal fluid by insertion of a hypodermic needle through the lumbar region of the spine, into the space surrounding the spinal cord and nerves

Mass fragmentography   Quantitative analysis of compounds by measurement of specific fragments using mass spectrometry

Mass spectrometry   Analysis of the chemical structure of a compound by measurement of the molecular weight of fragments of it. Fragments are formed by bombardment of the molecule by ions

Medulla oblongata   Area of brain lying below the pons

Mesencephalon   Area of the brain also known as the midbrain which contains the tegmentum and substantia nigra

Mesolimbic forebrain   The area containing the nucleus accumbens, olfactory tubercle, and projections to cortex

Microsomes   Small subcellular particles involved in metabolism

Migraine   A syndrome characterized by localized headache and often accompanied by nausea, vomiting, and sensory disturbances

Mitochondria   Rod shaped subcellular organelles involved in metabolism

Monoamine   Generic name for the catecholamine or indoleamine neurotransmitters

Monozygotic   In twin studies, refers to twins who have developed from a single ovum and therefore have the same genetic characteristics

Neuromodulator   Term applied to a compound postulated to regulate the action of a neurotransmitter. Particularly used to describe certain peptides (see Chapter 2, Part 5)

Neuromuscular block   Interruption of the transmission of nervous impulse to the muscle by preventing the action of the transmitter at the receptor on the muscle surface

Neuroregulator   Term sometimes used to describe compounds that have not been shown to fulfil all the criteria of a neurotransmitter (see Section 2.21)

Neurotic   Refers to psychiatric conditions where contact with reality and insight are maintained. Often used loosely and ambiguously

Nigrostriatal pathway   The neural projection from cell bodies in the substantia nigra to the striatum

Nominal scales   A rating scale which involves only the identification and not quantification of characteristics, which are not hierarchical, e.g. freezing, not freezing

Nuclear schizophrenia   Refers to the core clinical symptomatology of schizophrenia rather than associated or social factors

Oedema   Swelling due to the presence of excess fluid in the intercellular spaces of the body. American spelling: edema

Ordinal scales   A rating scale which involves the allocation of characteristics to categories which have a hierarchical relationship, e.g. hot, warm, cold

Palpitations   Unduly rapid beating of the heart which is noticed by the subject

Paper chromatography   Separation of compounds by placing a dried mixture on a filter paper and allowing a solvent to move across the paper by capillary action. Separation of compounds up the paper depends on their relative solubility in the solvent

Paranoia   The state of having delusions commonly but not necessarily persecutory

Particulate   Usually applied to that fraction of a tissue homogenate which contains the subcellular particles

Passivity feelings   Feelings of being under the control or will of an outside agency

Penetrance   In genetics, the degree to which a genetically inherited characteristic is expressed

Perseverative   A persistent reference to a theme, persistent use of a word or phrase, or a persistent behaviour which is out of context

Phobia   A persistent abnormal fear of some situation or object

Phospholipid   A lipid containing phosphorus

Pineal   A small gland in the midline of the brain. In many animals it responds to diurnal changes in light. Site of synthesis of metatonin (5-methoxy-N-acetyl serotonin)

Plasma   Blood from which the cells have been removed, without the blood being allowed to clot

Platelet   Small blood constituents, approx 2–3 nm in diameter, which are formed from larger precursor cells, and are involved in blood clotting. Also called thrombocytes

Polydipsia   Excessive drinking

Polyuria   The passage of large quantities of urine

Pons   Area of the hindbrain lying beneath the cerebellum

Post-mortem   Autopsy

Precursor   A compound which is metabolized to another compound. Usually used in reference to compounds which are metabolized to neurotransmitters

Presynaptic   Refers to events or structures proximal to the synapse

Process schizophrenia   Refers to core clinical symptomatology of schizophrenia rather than associated or social factors

Psychotic   Refers to psychiatric conditions where contact with reality and insight are lost. Often used loosely and ambiguously

Psychotropic drug   A drug which acts on psychic mood behaviour or experience

Putamen   Area of the brain within the corpus striatum

Radioimmunoassay   An assay technique which uses an antibody to the compound (X) to be measured. The displacement of a fixed concentration of radiolabelled compound X provides a measure of the unlabelled compound present

Radiolabelled compound Compound synthesized to contain one or more radioactive atoms at particular parts of the molecule

Rapid eye movement (REM) sleep Stage of sleep where there is high brain activity and rapid eye movements. It is associated with dreaming

Receptor A membrane site to which compounds specifically bind. See Section 2.14

Receptor blocker Archaic and rather misleading phrase denoting a receptor antagonist

Reliability A measure of the similarity of scores produced by rating instruments when used under similar but slightly different conditions, e.g. test-retest reliability, inter-rater reliability. See also sensitivity and specificity

Rhinorrhoea The persistent discharge of thin nasal mucus

Schiff's base Formation of an aldimine by condensation of an aliphatic or aromatic amine with an aldehyde

Seizure Uncontrolled or paroxysmal brain activity which may be expressed through the motor system

Sensitivity 1. Measure of the ability of a rating instrument to identify all those items of relevance in a study population. 2. Pharmacological and behavioural; see subsensitivity and supersensitivity

Serum Fluid formed from blood as a result of clotting

Specificity 1. Measure of the ability of a rating instrument to identify only those items of relevance in a study population. 2. Sometimes used in discussion of the characteristics of receptor or enzymes to recognize defined chemical structures

Stereotaxic surgery A method of producing accurately placed lesions in the brain, often by electrocoagulation or by implantation of radioactive pellets

Stereotypy The persistent repetition of particular movements

Striatum See corpus striatum

Subsensitivity 1. Pharmacological: decreased response to a fixed concentration of drug; shift of log dose-response to right. 2. Behavioural: decreased behavioural response to fixed dose of drug

Substantia nigra Region of midbrain containing pigmented cell bodies

Supersensitivity 1. Pharmacological: increased response to a fixed concentration of drug; shift of the log dose-response curve to left. 2. Behavioural: increased behavioural response to a fixed dose of drug

Synaptic cleft The space between the terminal process of one neuron and the receptor site of another

Synaptosomes The pinched off and resealed nerve endings formed by careful homogenization of brain tissue in isotonic saline

Tachycardia Rapid beating of the heart

Therapeutic index The ratio between the dose of a drug needed to produce a therapeutic effect (taken as unity) and the dose of the drug which is toxic

Tuberoinfundibular system The system connecting the hypothalamus with the pituitary

Up-regulation   An adaptation, usually of a receptor, to increase the function of the system

Validity   Overall assessment of the sensitivity and specificity of a rating instrument

Vas deferens   Duct which runs from the testis joining with other ducts to open into the urethra

Vasoconstriction   Reduction in the diameter of blood vessels by contraction of the circular muscles in their walls

Ventral tegmental area   Area of the midbrain dorsal to the substantia nigra

Ventricles   Cavities within the brain which contain cerebrospinal fluid

Volume of distribution   The apparent volume of the body through which a drug is distributed were it present throughout at the concentration found in plasma

References

Adams, P. W., Rose, D. P., Folkard, J., Wynn, V., Seed, M. and Strong, R. (1973) Effect of pyridoxine hydrochloride (Vitamin B_6) upon depression associated with oral contraceptives. *Lancet* i, 897

Aghajanian, G. K. and Haigler, H. J. (1975) Hallucinogenic indoleamines: preferential action upon presynaptic serotonin receptors. *Psychopharmac Commun* 1, 619

Akagi, H., Green, A. R. and Heal, D. J. (1981) Repeated electroconvulsive shock attenuates clonidine-induced hypoactivity in both mice and rats. *Br J Pharmac* 73, 230

Åsberg, M., Cronholm, B., Sjöqvist, F. and Tuck, D. (1971) Relationship between plasma level and therapeutic effect of nortriptyline. *Br Med J* 3, 331

Åsberg, M., Thoren, P., Träskman, L., Bertilsson, L. and Ringberger, V-A. (1976) 'Serotonin depression' – A biochemical subgroup within the affective disorders. *Science* 191, 478

Åsberg, M., Träskman, L. and Thoren, P. (1976) 5-HIAA in the cerebrospinal fluid – A biochemical suicide predictor. *Arch gen Psychiat* 33, 1193

Ashcroft, G. W., Crawford, T. B. B., Eccleston, D., Sharman, D. F., McDougall, E. J., Stanton, J. S. and Binns, J. K. (1966) 5-Hydroxyindole compounds in the cerebrospinal fluid of patients with psychiatric or neurological diseases. *Lancet* ii, 1049

Balldin, J., Eden, S. *et al.* (1980) Electroconvulsive therapy in Parkinson's syndrome with 'on-off' phenomenon. *J Neural Transm* 47, 11

Barbeau, A. (1979) Role of peptides in the pathogenesis and treatment of Parkinson's disease. In: *Central Nervous System Effects of Hypothalamic Hormones and other Peptides*, p. 403 (Eds Collu, R., Barbeau, A., Ducharme, J. and Rochefort, J.) Raven Press, New York

Barber, R. and Saito, K. (1976) Light microscope visualisation of GAD and GABA-T in immunocytochemical preparations of rodent CNS. In: *GABA in Nervous System Function*, p 113 (Eds Roberts, E., Chase, T. N. and Tower, D. B.) Raven Press, New York

Barchas, J. D., Berger, P. A., Ciaranello, R. D. and Elliot, G. R. (1977) *Psychopharmacology: from Theory to Practice* Oxford Univ. Press, New York

Bartholini, G. and Pletscher, A. (1967) Cerebral accumulation and metabolism of ^{14}C dopa after selective inhibition of peripheral decarboxylase. *J Pharmac exp Ther* 161, 14

Baumann, P. A. and Maitre, L. (1977) Blockade of presynaptic receptors and of amine uptake in the rat by the antidepressant mianserin. *Naunyn-Schmiedebergs Arch Pharmac* 300, 31

Beaumont, A. and Hughes, J. (1979) Biology of opioid peptides. *Ann Rev Pharmac Toxic* 19, 245

Bergstrom, D. A. and Kellar, K. J. (1979) Effect of electroconvulsive shock on monoaminergic receptor binding sites in rat brain. *Nature* 278, 464

Bernheimer, H. and Hornykiewicz, O. (1973) Brain amines in Huntington's chorea. *Adv Neurol* **1**, 525

Bertilsson, L. and Costa, E. (1976) Mass fragmentographic quantitation of glutamic acid and γ-aminobutyric acid in cerebellar nuclei and sympathetic ganglia of rats. *J Chromat* **118**, 395

Beskow, J., Gottfries, C. G., Roos, B-E. and Winblad, B. (1976) Determination of monoamines and monoamine metabolites in the human brain: post mortem studies in a group of suicides and in a control group. *Acta Psychiat Scand* **53**, 7

Bird, E. D. and Iversen, L. L. (1974) Huntington's chorea – post mortem measurement of glutamic acid decarboxylase, choline acetyltransferase and dopamine in basal ganglia. *Brain* **97**, 457

Bird, E. D. and Iversen, L. L. (1977) Neurochemical findings in Huntington's chorea. In: *Essays Neurochem Neuropharmac* **1**, 177 (Eds Youdim, M. B. H., Lovenberg, W., Sharman, D. F. and Lagnado, J. R.) John Wiley & Sons, Chichester

Bird, E. D., Mackay, A. V. P. *et al.* (1973) Reduced glutamic acid decarboxylase activity of post-mortem brain in Huntington's chorea. *Lancet* **i**, 1090

Bird, E. D., Spokes, E. G., Barnes, J., Mackay, A. V. P., Iversen, L. L. and Shepherd, M. (1977) Increased brain dopamine and reduced glutamic acid decarboxylase and choline acetyltransferase activity in schizophrenia and related psychoses. *Lancet* **ii**, 1157

Birkmayer, W., Riederer, P., Ambrozi, L. and Youdim, M. B. H. (1977) Implications of combined treatment with Madopar and L-deprenil in Parkinson's disease. A long term study. *Lancet* **i**, 439

Blaustein, M. P. and Ecton, A. C. (1975) Barbiturate inhibition of calcium uptake by depolarized nerve terminals *in vitro*. *Mol Pharmac* **11**, 369

Bleuler, E. (1911) *Dementia Praecox or the Group of Schizophrenics*. Vienna (translated J. Kemkin, 1950) Int. Univ. Press, New York

Bochner, F., Carruthers, G., Kampmann, J. and Steiner, J. (1978) *Handbook of Clinical Pharmacology* Little, Brown, Boston

Bonn, J., Turner, P. and Hicks, D. C. (1972) Beta-adrenergic-receptor blockade with practolol in treatment of anxiety. *Lancet* **i**, 814

Bowen, D. M. (1980) Biochemical evidence for selective vulnerability in Alzheimer's disease. In: *Biochemistry of Dementia*, p 77 (Ed Roberts, P. J.) John Wiley & Sons, Chichester

Bowen, D. M., Smith, C. B., White, P. and Davidson, A. N. (1976) Neurotransmitter related enzymes and indices of hypoxia in senile dementia and other bioatrophies. *Brain* **99**, 459

Bowen, D. M., White, P., *et al.* (1979) Accelerated ageing or elective neuronal loss as an important cause of dementia? *Lancet* **i**, 11

Bradbury, A. F., Smyth, D. G., Snell, C. K., Birdsall, N. J. M. and Hulme, F. C. (1976) C-fragment of lipotropin has a high affinity for brain opiate receptors. *Nature* **260**, 793

Braestrup, C., Nielsen, M. and Olsen, C. E. (1980) Urinary and brain β-carboline-3-carboxylates as potent inhibitors of brain benzodiazepine receptors. *Proc Nat Acad Sci* **77**, 2288

Briley, M., Langer, S. Z., Raisman, R., Sechter, D. and Zarifian, E. (1980) Tritiated imipramine binding sites are decreased in platelets of untreated depressed patients. *Science* **209**, 303

Brockington, I. F., Kendell, R. E. and Leff, J. P. (1978) Definitions of schizophrenia: concordance and prediction of outcome. *Psychol Med* **8**, 837

Brown, G. W. and Harris, T. O. (1978) *Social origins of depression*. London, Tavistock Publications

Bulat, M. and Zivković, B. (1971) Origin of 5-hydroxyindole acetic acid in the spinal fluid. *Science* **173**, 738

Bunney, W. E. and Post, R. M. (1977) Catecholamine agonist and receptor hypothesis of affective illness: paradoxical drug effects. In: *Neuroregulators and Psychiatric Disorders*, p 151 (Eds Usdin, E., Hamburg, D. A. and Barchas, J. D.) Oxford Univ Press New York

Burgus, R., Dunn, T. E. *et al.* (1970) Characterisation of ovine hypothalamic hypophysiotropic TSH-releasing factor. *Nature* **226**, 321

Cade, J. F. J. (1949) Lithium salts in the treatment of psychotic excitement. *Med J Australia* **36**, 349

Carlsson, A. and Lindqvist, M. (1963) Effect of chlorpromazine and haloperidol on formation of 3-methytyramine and normetanephrine in mouse brain. *Acta pharmac toxicol* **20**, 140

Cerletti, U. and Bini, L. (1938) Un nuevo metodo di shockterapie 'L'elettroshock'. *Boll Acad Med Roma* **64**, 136

Chang, H. C. and Gaddum, J. H. (1933) Choline esters in tissue extracts. *J Physiol (Lond)* **79**, 255

Chang, M. M., Leeman, S. E. and Niall, H. D. (1971) Amino acid sequence of substance P. *Nature (New Biol)* **232**, 82

Clow, A., Jenner, P. and Marsden, C. D. (1979) Changes in dopamine-mediated behaviour during one year's neuroleptic administration. *Eur J Pharmac* **57**, 365

Cochran, E., Robins, E. and Grote, S. (1976) Regional serotonin levels in brain: comparison of depressive suicides and alcoholic suicides with controls. *Biol Psychiat* **11**, 283

Connell, P. H. (1958) *Amphetamine Psychosis*. Chapman & Hall, London.

Cooper, J. R., Bloom, F. E. and Roth, R. H. (1974) *The Biochemical Basis of Neuropharmacology*, 2nd Edition. Oxford Univ Press, New York

Cooper, J. R., Bloom, F. E. and Roth, R. H. (1978) *The Biochemical Basis of Neuropharmacology*, 3rd Edition. Oxford Univ Press, New York

Coppen, A. (1973) Role of Serotonin in affective disorders. In: *Serotonin and Behaviour*, p 523 (Eds Barchas, J. and Usdin, E.) Academic Press, New York

Coppen, A., Eccleston, E. G. and Peet, M. (1973) Total and free tryptophan concentration in plasma of depressive patients. *Lancet* **ii**, 60

Coppen, A., Ghose, K., *et al.* (1978) Amitriptyline plasma concentration and clinical effect. *Lancet* **i**, 63

Coppen, A., Shaw, D. M. and Farrell, J. P. (1963) Potentiation of the antidepressive effect of a monoamine oxidase inhibitor by tryptophan. *Lancet* **i**, 79

Costa, E. (1970) Simple neuronal models to estimate turnover rate of noradrenergic transmitters *in vivo. Adv Biochem Psychopharmac* **2**, 169

Costa, E. (1979) The role of gamma-aminobutyric acid in the action of 1,4 benzodiazepines. *Trends Pharmac Sci* **1**, 41

Costa, E. and Sandler, M. (1972) Monoamine oxidase – new vistas. *Adv Biochem Psychopharmac* **5**, Raven Press, New York

Costall, B. and Naylor, R. J. (1977) A comparison of the abilities of typical neuroleptic agents and of thioridizine, clozapine, sulpiride and metoclopramide to antagonise the hyperactivity induced by dopamine applied intracerebrally to areas of the extrapyramidal and mesolimbic systems. *Eur J Pharmac* **40**, 9

Cott, J. and Engel, J. (1977) Suppression by GABAergic drugs of the locomotor stimulation induced by morphine, amphetamine and apomorphine: evidence for both pre- and post-synaptic inhibition of catecholamine systems. *J Neural Trans* **40**, 263

Cotzias, G. C., Papavasiliov, P. S. and Gellene, R. (1969) Modification of parkinsonism: chronic treatment with L-Dopa. *New Engl J Med* **280**, 337

Cotzias, G. C., Van Woert, M. H. and Schiffer, L. M. (1967) Aromatic amino acids and modification of parkinsonism. *New Engl J Med* **276**, 374

Cowen, P. J., Green, A. R., Martin, I. L. and Nutt, D. J. (1981) Ethyl β-carboline carboxylate lowers seizure threshold and antagonises flurazepam-induced sedation in rats. *Nature* **290**, 54

Creese, I., Burt, D. R. and Snyder, S. H. (1977) Dopamine receptor binding enhancement accompanies lesion induced behavioral supersensitivity. *Science* **197**, 596

Crews, F. T. and Smith, C. B. (1978) Presynaptic alpha-receptor supersensitivity after long term antidepressant treatment. *Science* **202**, 32

Cross, A. J., Crow, T. J., Perry, E. K., Perry, R. H., Blessed, G. and Tomlinson, B. E. (1981) Reduced dopamine β-hydroxylase activity in Alzheimer's disease. *Br Med J* **282**, 93

Crow, T. J., Baker, H. F. *et al.* (1979a) Monoamine mechanisms in chronic schizophrenia: post mortem neurochemical findings. *Br J Psychiat* **134**, 249

Crow, T. J., Deakin, J. F. W. and Longden, A. (1975) Do antipsychotics act by dopamine receptor blockade in the nucleus accumbens? *Br J Pharmac* **55**, 295

Crow, T. J., Ferrier, I. N. *et al.* (1979b) Characteristics of patients with schizophrenia or neurological disorder and virus-like agent in cerebrospinal fluid. *Lancet* **i**, 842

Crow, T. J., Johnstone, E. C. and McClelland, H. A. (1976) The coincidence of schizophrenia and parkinsonism: some neurochemical implications. *Psychol Med* **6**, 227

Crow, T. J., Johnstone, E. C. and Owen, F. (1979) Research on schizophrenia In: *Recent Advances in Clinical Psychiatry* **3**, 1 (Ed Granville-Grossman, K.) Churchill-Livingstone, Edinburgh

Cuatrecasas, P. (1974) Membrane receptors. *Ann Rev Biochem* **43**, 169

Cuello, A. C. and Kanazawa, I. (1978) The distribution of substance P immunoreactive fibers in the rat central nervous system. *J comp Neurol* **178**, 129

Curzon, G. (1969) Tryptophan pyrrolase – a biochemical factor in depressive illness. *Br J Psychiat* **115**, 1367

Curzon, G., Fernando, J. C. R. and Marsden, C. A. (1978) 5-Hydroxytryptamine: the effects of impaired synthesis on its metabolism and release in rat. *Br J Pharmac* **63**, 627

Curzon, G., Gumpert, E. J. W. and Sharpe, D. M. (1971) Amine metabolites in the lumbar cerebrospinal fluid of humans with restricted flow of cerebrospinal fluid. *Nature (New Biol)* **231**, 189

Curzon, G., Kantamanini, B. D., Van Boxel, P., Gillman, P. K., Bartlett, J. R. and Bridges, P. K. (1980) Substances related to 5-hydroxytryptamine in plasma and in lumbar and ventricular fluids of psychiatric patients. *Acta psychiat Scand* **Suppl 280**, 3

Davies, D. and Verth, A. H. (1978) Regional distribution of muscarinic acetylcholine receptors in normal and Alzheimer-type dementia brains. *Brain Res* **138**, 385

Davis, K. L. and Yamamura, H. I. (1978) Cholinergic underactivity in human memory disorders. *Life Sci* **23**, 1729

Dawbarn, D., Long, S. K. and Pycock, C. J. (1981) Increased 5-hydroxytryptamine receptor mechanisms in rats after chronic neuroleptic treatment. *Br J. Pharmac* **73**, 149

Dayan, A. D. (1974) The brain, ageing and dementia. *Psychol Med* **4**, 349

Deakin, J. F. W. and Dostrovsky, J. O. (1978) Involvement of the periaqueductal grey matter and spinal 5-hydroxytryptaminergic pathways in morphine analgesia: effects of lesions and 5-hydroxytryptamine depletion. *Br J Pharmac* **63**, 159

Deitrich, R. and Erwin, V. (1980) Biogenic amine–aldehyde condensation products: tetrahydroisoquinolines and triptolines (β-carbolines). *Ann Rev Pharmac Toxic* **20**, 55

Delay, J., Deniker, P. and Hart, J. M. (1952) Utilisation en therapeutique psychiatrique d'une phenothiazine d'action centrale élective. *Ann méd-psychol (Paris)* **110**, 112

Deutsch, J. A. (1973) The cholinergic synapse and the site of memory. In: *The Physiological Basis of Memory* p 59 (Ed: Deutsch, J. A.) Academic Press, New York

De Wied, D., Van Wimersma Greidanus, T. B., Bohus, B., Urban, I and Gispen, W. H. (1976) Vasopressin and memory consolidation. *Progr Brain Res* **45**, 181

De Wied, D., Witter, A. and Greven, H. M. (1975) Behaviourally active ACTH analogues. *Biochem Pharmac* **24**, 1463

Drew, W. G. and Miller, L. L. (1974) Cannabis: neural mechanisms and behaviour – a theoretical review. *Pharmacology* **11**, 12

Ebstein, R. P., Eliashar, S. and Belmaker, R. H. (1980) The effect of chronic lithium on adenylate cyclase and dopamine-mediated animal behaviours. In: *Enzymes and Neurotransmitters in Mental Disease* p 395 (Eds Usdin, E., Sourkes, T. L. and Youdim, M. B. H.) John Wiley & Sons, Chichester

Eccleston, D. and Nicholaou, N. (1978) The influence of L-tryptophan and monoamine oxidase inhibitors on catecholamine metabolism in rat brain. *Br J Pharmac* **64**, 341

Ehringer, H. and Hornykiewicz, O. (1960) Verteilung von noradrenalin und dopamin (3-hydroxytyramin) im gehirn des extrapyramidalen systems. *Kline Wochenschr* **38**, 1236

Elliot, G. R. and Hollman, R. B. (1977) Tryptolines as potential modulators of serotonergic function. In: *Neuroregulators and Psychiatric Disorders*, p 220 (Eds Usdin, E., Hamburg, D. A. and Barchas, J. D.) Oxford Univ Press, New York

Enna, S. J., Bird, E. D. *et al.* (1976) Huntington's chorea: changes in neurotransmitter receptors in the brain. *New Engl J Med* **294**, 1305

Feighner, J. P., Robins, E., Guze, S. B., Woodruff, R. A., Winokur, G. and Munoz, R. (1972) Diagnostic criteria for use in psychiatric research. *Arch Gen Psychiat* **26**, 57

Fernstrom, J. D. and Wurtman, R. J. (1974) Effects of the diet on brain serotonin. *Sci Am* **230**, 84

Fink, M. (1974) Induced seizures and human behavior. In: *Psychobiology of Convulsive Therapy* p 1 (Eds Fink, M., Kety, S., McGaugh, J. and Williams, T. A.) V. H. Winston & Sons, Washington D.C.

Fink, M. (1979) *Convulsive Therapy: Theory and Practice*. Raven Press, New York

Forchetti, C. M. and Meek, J. L. (1981) Evidence for a tonic GABA-ergic control of serotonin neurones in the median raphe nucleus. *Brain Res* **206**, 208

Freedman, A. M., Kaplan, H. I. and Sadock, B. J. (1980) *Comprehensive Textbook of Psychiatry* 3rd Edition (3 vols) Williams & Wilkins, Baltimore

Friedhoff, A. and Van Winkle, E. (1962) The characteristics of an amine found in the urine of schizophrenic patients. *J Nerv Ment Dis* **135**, 550

Gaddum, J. H. (1953) Antagonism between lysergic acid diethylamide and 5-hydroxy-tryptamine. *J Physiol* **121**, 15

Gale, K. (1980) Chronic blockade of dopamine receptors by antischizophrenic drugs enhances GABA binding in substantia nigra *Nature* **283**, 569

Gamble, S. J. and Waddington, J. L. (1981) Six months phenothiazine treatment differentially influences distinct dopamine-mediated behaviours. *Br J Pharmac* **73**, 240

Garattini, S., Pujol, J. F. and Samanin, R. (1978) *Interactions between Putative Neurotransmitters in the Brain*. Raven Press, New York

Garelis, E., Young, S. N., Lal, S. and Sourkes, T. L. (1974) Monoamine metabolites in lumbar CSF: the question of their origin in relation to clinical studies. *Brian Res* **79**, 1

Gelder, M. G., Gath, D. H. and Mayou, R. A. (1982) *The Oxford Textbook of Psychiatry* (In Press) Oxford Univ Press, Oxford

Gershon, S. and Shopsin, B. (1973) *Lithium: its Role in Psychiatric Research and Treatment*. Plenum Press, New York

Glaser, G. H., Penry, J. K. and Woodbury, D. M. (1980) Antiepileptic drugs, mechanisms of action. *Adv Neurol* **27** Raven Press, New York

Glover, V., Sandler, M., Owen, F. and Riley, G. J. (1977) Dopamine is a monoamine B substrate in man. *Nature* **265**, 80

Gottfries, C. G. (1980) Amine metabolism in normal ageing and in dementing disorders. In: *Biochemistry of Dementia* p 213 (Ed Roberts, P. J.) J Wiley & Sons, Chichester

Gottfries, C. G., Adolfsson, R., Oreland, L., Roos, B-E. and Winblad, B. (1979) Monoamines and their metabolites and monoamine oxidase activity related to age and some dementia disorders. In: *Drugs and the Elderly. Perspectives in Geriatric Clinical Pharmacology* p 189 (Eds Crook, J. and Stevenson, I. H.) Macmillan, London

Gottfries, C. G., Gottfries, I. and Roos, B-E. (1968) Disturbances of monoamine metabolism in the brains from patients with dementia senilis and M6 Alzheimer *Excerpta Med Int Cong Ser* **180**, 310

Gottfries, C. G., Gottfries, I. and Roos, B-E. (1969) The investigation of homovanillic acid in the human brain and its correlation to senile dementia. *Br J Psychiat* **115**, 563

Grahame-Smith, D. G. (1971) Studies *in vivo* on the relationship between brain tryptophan, brain 5-HT synthesis and hyperactivity in rats treated with a monoamine oxidase inhibitor and L-tryptophan. *J Neurochem* **18**, 1053

Grahame-Smith, D. G., Green, A. R. and Costain, D. W. (1978) The mechanism of the antidepressant action of ECT. *Lancet* **i**, 254

Granville-Grossman, K. L. and Turner, P. (1966) The effect of propranolol on anxiety. *Lancet* **i**, 788

Green, A. R. (1978) The effects of dietary tryptophan and its peripheral metabolism on brain 5-hydroxytryptamine synthesis and function. In: *Essays in Neurochem and Neuropharmac* **3**, 103, John Wiley & Sons, Chichester

Green, A. R. (1980) The behavioural and biochemical consequences of repeated electroconvulsive shock administration to rats and the possible clinical relevance of these changes. In: *Enzymes and Neurotransmitters in Mental Disease* p 455 (Eds Usdin, E., Sourkes, T. L. and Youdim, M. B. H.) John Wiley & Sons, Chichester

Green, A. R., Aronson, J. K., Curzon, G. and Woods, H. F. (1980) Metabolism of an oral tryptophan load. II Effect of pretreatment with the putative tryptophan pyrrolase inhibitors nicotinamide or allopurinol. *Br J Clin Pharmac* **10**, 611

Green, A. R., Bloomfield, M. R., Woods, H. F. and Seed, M. (1978) Metabolism of an oral tryptophan load by women and evidence against the induction of tryptophan pyrrolase by oral contraceptives. *Br J Clin Pharmac* **5**, 233

Green, A. R. and Costain, D. W. (1979) The biochemistry of depression. In: *Psychopharmacology of Affective Disorders* p 14 (Eds Paykel, E. S. and Coppen, A.) Oxford Univ Press, Oxford

Green, A. R. and Deakin, J. F. W. (1980) Brain noradrenaline depletion prevents ECS-induced enhancement of serotonin- and dopamine-mediated behaviour. *Nature* **285**, 232

Green, A. R. and Grahame-Smith, D. G. (1975) 5-Hydroxytryptamine and other indoles in the central nervous system. In: *Handbook of Psychopharmacology* (Eds Iversen, L. L., Iversen, S. D. and Snyder, S. H.) **3**, 169 Plenum Press, New York

Green, A. R. and Grahame-Smith, D. G. (1976) (−)-Propranolol inhibits the behavioural responses of rats to increased 5-hydroxytryptamine in the central nervous system. *Nature* **262**, 594

Green, A. R. and Kelly, P. H. (1976) Evidence concerning the involvement of 5-hydroxytryptamine in the locomotor activity produced by amphetamines or tranylcypromine plus L-dopa. *Br J Pharmac* **57**, 141

Green, A. R. and Youdim, M. B. H. (1975) Effects of monoamine oxidase inhibitor by clorgyline, deprenil or tranycypromine on 5-hydroxytryptamine concentrations in rat brain and hyperactivity following subsequent tryptophan administration. *Br J Pharmac* **55**, 415

Greengard, P. (1976) Possible role for cyclic nucleotides and phosphorylated membrane proteins in post-synaptic actions of neurotransmitters. *Nature* **260**, 101

Guidotti, A. (1978) Synaptic mechanisms in the action of benzodiazepines. In: *Psychopharmacology, a Generation of Progress*, p 1349 (Eds Lipton, M. A., Di Mascio, A. and Killam, K. F.) Raven Press, New York

Guidotti, A., Toffano, G. and Costa, E. (1978) An endogenous protein modulates the affinity of GABA and benzodiazepine receptors in rat brain. *Nature* **275**, 553

Hamilton, M. (1960) A rating scale for depression. *J Neurol Neurosurg Psychiat* **23**, 56

Heal, D. J. and Green, A. R. (1979) Administration of thyrotropin releasing hormone (TRH) to rats releases dopamine in n. accumbens but not n. caudatus. *Neuropharmacology* **18**, 23

Henry, J. L. (1977) Substance P and pain: a possible relation in afferent transmission. In: *Substance P*, p 231 (Eds Von Euler, U. S. and Pernow, B.) Raven Press, New York

Hiley, C. R. and Bird, E. D. (1974) Decreased muscarinic receptor concentration in post mortem brain in Huntington's chorea. *Brain Res* **80**, 355

Hill, P., Murray, R. and Thorley, A. (1979) *Essentials of Postgraduate Psychiatry.* Academic Press, London.

Hillestad, L., Hansen, T. and Melsom, H. (1974) Diazepam metabolism in normal man. Serum concentration and diurnal effect after oral administration and cumulation. *Clin Pharmac Ther* **16**, 485

Hökfelt, T., Johansson, O., Ljungdahl, A., Lundberg, J. M. and Schultzberg, M. (1980) Peptidergic neurones. *Nature* **284**, 515

Hökfelt, T., Ljungdahl, A. *et al.* (1977) Some attempts to explore possible central Gabergic mechanisms with special reference to control of dopamine neurons. In *Neuroregulators and Psychiatric Disorders* p 358 (Eds Usdin, E., Hamburg, D. A. and Barchas, J. D.) Oxford Univ Press, New York

Hollister, L. E. (1977) Some general thoughts about endogenous psychotogens. In: *Neuroregulators and Psychiatric Disorders* p 550 (Eds Usdin, E., Hamburg, D. A. and Barchas, J. D.) Oxford Univ Press, New York

Hollister, L., Motzenbecker, F. P. and Degan, R. O. (1961) Withdrawal reactions from chlordiazepoxide (Librium). *Psychopharmac* **2**, 63

Hong, J. S., Yang, H-Y. T., Fratta, W. and Costa, E. (1978) Rat striatal methionine-enkephalin content after chronic treatment with cataleptogenic and non-cataleptogenic antischizophrenic drugs. *J Pharmac exp Ther* **205**, 141

Hornykiewicz, O. (1966) Dopamine (3-hydroxytyramine) and brain function. *Pharmac Rev* **18**, 925

Hornykiewicz, O. (1977) Psychopharmacological implications of dopamine and dopamine antagonists: a critical evaluation of current evidence. *Pharmac Rev* **17**, 545

Hornykiewicz, O. (1973) Dopamine in the basal ganglia: its role and therapeutic implications (including the clinical use of L-Dopa). *Br Med Bull* **29**, 172

Hughes, J., Smith, T. W., Kosterlitz, H. W., Fothergill, L. A., Morgan, B. A. and Morris, H. R. (1975) Identification of two related pentapeptides from the brain with potent opiate agonist activity. *Nature* **258**, 577

Itil, T. M., Polvan, N. and Hsu, W. (1972) Clinical and EEG effects of GB-94 a tetracyclic antidepressant (EEG model) in discovery of a new psychotropic drug. *Curr Ther Res* **14**, 395

Iversen, L. L. (1975a) Dopamine receptors in the brain. *Science* **188**, 1084

Iversen, L. L. (1975b) Uptake processes for biogenic amines. In: *Handbook of Psychopharmacology* (Eds Iversen, L. L., Iversen, S. D. and Snyder, S. H.) **3**, 381 Plenum Press, New York

Iversen, L. L. (1977) Anti-anxiety receptors in the brain? *Nature* **266**, 678

Iversen, L. L., Iversen, S. D. and Snyder, S. H. (1975) *Handbook of Psychopharmacology* **3**, Plenum Press, New York

Iversen, L. L. and Mackay, A. V. P. (1979) Pharmacodynamics of antidepressants and antimanic drugs. In: *Psychopharmacology of Affective Disorders* p 60 (Eds Paykel, E. S. and Coppen, A) Oxford Univ Press, Oxford

Jackson, I. M. D. and Reichlin, S. (1979) Distribution and biosynthesis of TRH in the nervous system. In: *Central Nervous System Effects of Hypothalamic Hormones and Other Peptides* p 3 (Eds Collu, R., Barbeau, A., Ducharme, J. R. and Rochefort, J. G.) Raven Press, New York

Jensen, K., Fruensgaard, K. *et al.* (1975) Tryptophan/imipramine in depression. *Lancet* **ii**, 920

Johnson, F. N. and Johnson, S. (1978) *Lithium in Medical Practice.* MTP Press, Lancaster

Johnson, K. M., Ho, B. T. and Dewey, W. L. (1976) Effects of Δ^9-tetrahydrocannabinol on neurotransmitter accumulation and release mechanism in rat forebrain synaptosomes. *Life Sci* **19**, 347

Johnston, J. P. (1968) Some observations upon a new inhibitor of monoamine oxidase in Brain tissue. *Biochem Pharmac* 17, 1285

Johnstone, E. C., Crow, T. J., Frith, C. D., Carney., M. W. P. and Price, J. S. (1978) Mechanism of the antipsychotic effect in the treatment of acute schizophrenia. *Lancet* i, 848

Joseph, M. H., Baker, H. F., Crow, T. J., Riley, G. J. and Risby, D. (1979) Brain tryptophan metabolism in schizophrenia: a post mortem study of metabolites on the serotonin and kynurenine pathways in schizophrenics and control subjects. *Psychopharmacology* 62, 279

Kanof, P. D. and Greengard, P. (1978) Brain histamine receptors as targets for antidepressant drugs. *Nature* 272, 329

Karczmar, A. G. (1975) Cholinergic influences on behaviour. In: *Cholinergic Mechanisms* p 501 (Ed Waser, P. G.) Raven Press, New York

Kebabian, J. W. and Calne, D. (1979) Multiple receptors for dopamine. *Nature* 277, 93

Kebabian, J. W., Petzold, G. L. and Greengard, P. (1972). Dopamine-sensitive adenylate cyclase in caudate nucleus of rat brain and its similarity to the 'dopamine receptor'. *Proc Nat Acad Sci* 69, 2145

Kelley, A. E., Stinus, L. and Iversen, S. D. (1979) Behavioural activation induced in the rat by substance P infusion into the ventral tegmental area: implication of dopaminergic A10 neurones. *Neurosci Lett* 11, 335

Kelly, P. H. and Iversen, L. L. (1975) LSD as an agonist at mesolimbic dopamine receptors *Psychopharmacologia* 45, 221

Kelly, P. H., Seviour, P. W. and Iversen, S. D. (1975) Amphetamine and apomorphine responses in the rat, to 6-OHDA lesions of the nucleus accumbens septi and corpus striatum. *Brain Res* 94, 507

Kendell, R. E. (1972) Schizophrenia. The remedy for diagnostic confusion. *Br J Hosp Med* 8, 383. Reprinted in: *Contemporary Psychiatry*, 1975 (Eds Silverstone, T. and Baraclough, B.) Headley Brothers, Ashford, Kent

Kerwin, R. W. and Pycock, C. (1979) Thyrotropin releasing hormone stimulates release of [³H] -dopamine from slices of rat nucleus accumbens *in vitro*. *Br J Pharmac* 67, 323

Kety, S. (1974) Biochemical and neurochemical effects of electroconvulsive shock. In: *Psychobiology of Convulsive Therapy* p 285 (Eds Fink, M., Kety, S., McGaugh, J. and Williams, T. A.) V. H. Winston & Sons, Washington D.C.

Klawans, H. L. (1973) The pharmacology of extrapyramidal movement disorders. *Monographs in Neurol Sci* 2, 64 Karger, Basle

Kosterlitz, H. W. (1980) Possible physiological roles of the enkephalins and endorphins. In: *Clinical Pharmacology and Therapeutics* p 33 (Ed Turner, P.) Macmillan, London

Kosterlitz, H. and Hughes, J. (1975) Some thoughts on the significance of enkephalin, the endogenous ligand. *Life Sci* 17, 91

Kosterlitz, H. W. and Waterfield, A. A. (1975) *In vitro* models in the study of structure – activity relationships of narcotic analgesics. *Ann Rev Pharmac* 15, 29

Kovacs, G. L., Bohus, B. and Versteog, D. G. G. (1979) The effects of vasopressin on memory process: the role of noradrenergic neurotransmission. *Neuroscience* 4, 1529

Kraepelin, E. (1919) *Dementia Praecox and Paraphrenia*. Translated by Barclay, R. M. Livingstone, Edinburgh.

Kragh-Sørenson, P., Eggert-Hansen, C., Baastrup, P. C. and Hindberg, E. F. (1976) Self inhibiting action of nortriptyline's antidepressive effect at high plasma level. *Psychopharmacologia* 45, 305

Kuriyama, K., Roberts, E. and Rubinstein, M. K. (1966) Elevation of γ-aminobutyric acid in brain with amino-oxyacetic acid and susceptibility to convulsive seizures in mice: a quantitative re-evaluation. *Biochem Pharmac* 15, 221

Lader, M. H. and Tyrer, P. J. (1972) Central and peripheral effects of propranolol and sotalol in normal human subjects. *Br J Pharmac* 45, 557

Langer, S. Z. (1977) Presynaptic receptors and their role in the regulation of transmitter release. *Br J Pharmac* **60**, 481

Lapin, I. P. and Oxenkrug, G. F. (1969) Intensification of the central serotonergic processes as a possible determinant of the thymoleptic effect. *Lancet* **i**, 132

Liljequist, S. (1978) Changes in the sensitivity of dopamine receptors in the nucleus accumbens and in the striatum induced by chronic ethanol administration. *Acta pharmac toxic* **43**, 19

Lindros, K. O. and Eriksson, C. J. P. (1975) *The Role of Acetaldehyde in the Actions of Ethanol*. Finnish Foundation for Alcohol Studies, Helsinki

Ljundahl, A., Hökfelt, T. and Nilsson, G. (1978) Distribution of substance P-like immunoreactivity in the central nervous system of the rat-1 cell bodies and nerve terminals. *Neuroscience* **3**, 861

Lloyd, K. G. (1978) Neurotransmitter interactions related to central dopamine neurons. *Essays in Neurochem and Neuropharmac* **3**, 129, John Wiley & Sons, Chichester

Lloyd, K. G. (1980) Indications for GABA neurons dysfunction in mental disease. In: *Enzymes and Neurotransmitters in Mental Disease* p 329 (Eds Usdin, E., Sourkes, T. L. and Youdim, M. B. H.) John Wiley & Sons, Chichester

Lloyd, K. G. and Davidson, L. (1979) Involvement of GABA neurons and receptors in Parkinson's Disease and Huntington's Chorea: A compensatory mechanism? *Adv Neurol* **24**, 293

Lloyd, K. G., Farley, I. J., Deck, J. H. N. and Hornykiewicz, O. (1974) Serotonin and 5-hydroxyindoleacetic acid in discrete areas of the brain stem of suicide victims and control patients. *Adv Biochem Psychopharmac* **11**, 387

Lord, J. A. H., Waterfield, A. A., Hughes, J. and Kosterlitz, H. W. (1977) Endogenous opioid peptides: multiple agonists and receptors. *Nature* **267**, 495

Mao, C. C., Peralta, E., Moroni, F. and Costa, E. (1978) The turnover rate of gamma-aminobutyric acid in the substantia nigra following electrical stimulation or lesioning of strionigral pathways. *Brain Res* **155**, 145

Marco, E., Mao, C. C., Cheney, D. L., Revuelta, A. and Costa, E. (1976) The effects of antipsychotics on the turnover rate of GABA and acetylcholine in rat brain nuclei. *Nature* **264**, 363

Marley, E. (1977) Monoamine oxidase inhibitors and drug interactions. In: *Drug Interactions* p 171 (Ed Grahame-Smith, D. G.) Macmillan, London

Marsden, C. A., Conti, J., Strope, E., Curzon, G. and Adams, R. N. (1979) Monitoring 5-hydroxytryptamine release in the brain of freely moving unanaesthetised rat using *in vivo* voltammetry. *Brain Res* **171**, 85

Marsden, C. D. (1976) Advances in the management of Parkinson's disease. *Scottish Med J* **21**, 139

Marsden, C. D., Tarsy, D. and Baldessarini, R. J. (1975) Spontaneous and drug induced movement disorders in psychotic patients. In: *Psychiatric Aspects of Neurological Disease* p 219 (Eds Benson, D. F. and Blumer, D.) Grune & Stratton Inc., New York

Martin, W. R., Eades, C. G., Thompson, J. A., Huppler, R. E. and Gilbert, P. E. (1976) The effects of morphine- and naloxone-like drugs in the nondependent and morphine-dependent chronic spinal dog. *J Pharmac exp Ther* **197**, 517

Matsson, B. (1974) *Clinical, Genetic and Pharmacological Studies in Huntington's Chorea* Umea University Medical Dissertations No 7 Umea, Sweden

Medical Research Council (1965) Report by the clinical psychiatry committee. Clinical trial of the treatment of depressive illness. *Br Med J* **1**, 881

Meldrum, B. S. (1975) Epilepsy and GABA-mediated inhibition. *Int Rev Neurobiol* **17**, 1

Mendlewicz, J. and Youdim, M. B. H. (1978) Antidepressant potentiation of 5-hydroxytryptophan by L-deprenyl a MAO type 'B' inhibitor. *J Neural Transm* **43**, 279

Middlemiss, D. N., Blakeborough, L. and Leather, S. R. (1977) Direct evidence for an interaction of β-adrenergic blockers with the 5-HT receptor. *Nature* **267**, 289

Miller, R. J. and Hiley, C. R. (1974) Antimuscarinic properties of neuroleptic drugs and drug induced parkinsonism. *Nature* **248**, 596

Mindham, R. H. S. (1979) Tricyclic antidepressants and amine precursors. In: *Psychopharmacology of Affective Disorders* p. 123 (Eds Paykel E. S. and Coppen, A.) Oxford Univ Press, Oxford

Neff, N. H., Gentleman, M., Parenti, M. and Commissiong, J. W. (1980) Dopaminergic neurons of spinal cord: a possible site for the action of neuroleptic drugs. In: *Enzymes and Neurotransmitters in Mental Disease* p. 37 (Eds Usdin, E; Sourkes, T. L. and Youdim, M. B. H.) John Wiley & Sons, Chichester

Neff, N. H. and Goridis, C. (1972) Neuronal monoamine oxidase: specific enzyme types and their rate of formulation. *Adv Biochem Psychopharmac* 5, 307 Raven Press, New York

Neff, N. H. and Tozer, T. N. (1968) *In vivo* measurement of brain serotonin turnover. *Adv Pharmac* 6A, 97

Neff, N. H., Tozer, T. N. and Brodie, B. B. (1967) Application of steady-state kinetics to studies of the transfer of 5-hydroxy indole acetic acid from brain to plasma. *J Pharmac Exp Ther* 158, 214

Osmond, H. and Smythies, J. (1952) Schizophrenia: a new approach. *J. Ment Sci* 98, 309

Overall, J. E. and Gorham, D. R. (1962) The Brief Psychiatric Rating Scale *Psychological Reports* 10, 799

Owen, F., Cross, A. J., Crow, T. J., Longden, A., Poulter, M. and Riley, G. J. (1978) Increased dopamine-receptor sensitivity in schizophrenia. *Lancet* ii, 223

Owen, F., Cross, A. J., Crow, T. J., Lofthouse, R. and Poulter, M. (1981) Neurotransmitter receptors in brains of schizophrenics. *Acta Psychiat Scand* 63, Suppl 291, 20

Pagel, J., Christian, S. T., Quale, E. S. and Monti, J. A. (1976) A serotonin sensitive adenylate cyclase in mature rat brain synaptic membranes. *Life Sci* 19, 819

Pare, C. M. B. (1975) Introduction to clinical aspects of monoamine oxidase inhibitors in the treatment of depression In: *Monoamine Oxidase and its Inhibition* p 271 Ciba Foundation Series 39 (Eds Wolstenholme, G. E. W. and Knight, J.) Elsevier, Amsterdam

Paton, W. D. M. (1975) Pharmacology of marijuana. *Ann Rev Pharmac* 15, 191

Paul, M. S., Syapin, P. J., Paugh, B. A., Moncada, V. and Skolnick, P. (1979) Correlation between benzodiazepine receptor occupation and anticonvulsant effects of diazepam. *Nature* 281, 688

Pearse, A. G. E. (1969) The cytochemistry and ultrastructure of polypeptide hormone producing cells of the APUD series and the embryological, physiologic and pathologic implication of the concept. *J Histochem Cytochem* 17, 303

Peroutka, S. J. and Snyder, S. H. (1979) Multiple serotonin receptors: differential binding of [^{3}H]-5-hydroxytryptamine, [^{3}H]-lysergic acid diethylamide and [^{3}H]spiroperidol. *Mol Pharmac* 16, 687

Perry, E. K. and Perry, R. H. (1980) The cholinergic system in Alzheimer's disease. In: *Biochemistry of Dementia* p 135 (Ed Roberts, P. J.) John Wiley & Sons, Chichester

Perry, T. L., Hansen, S. and Kloster, M. (1973) Huntington's Chorea: deficiency of γ-aminobutyric acid in brain. *New Engl J Med* 288, 337

Pert, C. B. and Snyder, S. H. (1973) Opiate receptor: demonstration in nervous tissue. *Science* 179, 1011

Peters, J. R., Elliot, J. M. and Grahame-Smith, D. G. (1979) Effect of oral contraceptives on platelet noradrenaline and 5-hydroxytryptamine receptors and aggregation. *Lancet* ii, 933

Pletscher, A. (1978) Platelets as models for monoaminergic neurons. *Essays in Neurochem and Neuropharmac* 3, 49 John Wiley & Sons, Chichester

Poirier, L. J. and Sourkes, T. L. (1965) Influence of the substantia nigra on the catecholamine content of the striatum. *Brain* 88, 181

Raisman, R., Briley, M. and Langer, S. Z. (1979) Specific tricyclic antidepressant binding sites in rat brain. *Nature* 281, 148 (1979)

Reisine, T. D., Yamamura, H. I., Bird, E. D., Spokes, E. and Enna, S. J. (1978) Pre- and post-synaptic neurochemical alterations in Alzheimer's disease. *Brain Res* **159**, 477

Richards, C. D. (1972) On the mechanism of barbiturate anaesthesia. *J. Physiol (Lond)* **227**, 749

Roberts, E., Chase, T. N. and Tower, D. B. (1976) *GABA in Nervous System Function.* Raven Press, New York

Roos, B-E. and Sjöstrom, R. (1969) 5-hydroxyindole acetic acid and homovanillic acid levels in the cerebrospinal fluid after probenecid application in patients with manic-depressive psychosis. *J Clin Pharmac* **1**, 153

Royal College of Psychiatrists (1977) The Royal College of Psychiatrists memorandum on the use of electroconvulsive therapy. *Br J Psychiat* **131**, 261

Rubin, R. P. (1970) The role of calcium in the release of neurotransmitter substances and hormones. *Pharmac Rev* **22**, 389

Sainsbury, P. (1968) Suicide and depression. In: *Recent Developments in Affective Disorders.* p 1 (Eds Coppen, A. and Walk, A.) Headley, Ashford Kent

Schally, A. V., Redding, T. W., Bowers, C. Y. and Barrett, J. F. (1969) Isolation and properties of porcine thyrotropin-releasing hormone. *J Biol Chem* **24**, 4077

Schildkraut, J. J. (1965) The catecholamine hypothesis of affective disorders: a review of the supporting evidence. *Am J Psychiat* **122**, 509

Schildkraut, J. J. (1973) Pharmacology – the effects of lithium on biogenic amines. In: *Lithium: its Role in Psychiatric Research and Treatment.* (Ed Gershon, S. and Schopsin, B.) Plenum Press, New York, London.

Schildkraut, J. J., Orsulak, P. J. *et al.* (1977) Recent studies on the role of catecholamines in the pathophysiology and classification of depressive disorders. In: *Neuroregulators and Psychiatric Disorders* p 122 (Eds Usdin, E., Hamburg, D. A. and Barchas, J. D.) Oxford Univ Press, New York

Schlesser, M. A., Winokur, G. and Sherman, B. M. (1979) Genetic subtypes of unipolar primary depressive illness distinguished by hypothalamic pituitary adrenal axis activity. *Lancet* **i**, 739

Schneider, K. (1959) *Klinische psychopathologie.* (Translated by Hamilton, M. W.) Grune & Stratton, New York

Schwartz, J-C. (1975) Histamine as a transmitter in brain. *Life Sci* **17**, 503

Sellinger-Barnette, M. M., Mendels, J. and Frazer, A. (1980) The effect of psychoactive drugs on beta-adrenergic receptor binding sites in rat brain. *Neuropharmacology* **19**, 447

Shopsin, B., Gershon, S., Goldstein, M., Friedman, E. and Wilk, S. (1975) Use of synthesis inhibitors in defining a role for biogenic amines during imipramine treatment in depressed patients. *Psychopharmac Commun* **1**, 239

Silver, A. (1974) *Biology of Cholinesterases.* North-Holland, Amsterdam

Simantov, R. and Snyder, S. H. (1976) Elevated levels of enkephalin in morphine-dependent rats. *Nature* **262**, 505

Sims, N. R., Smith, C. C. T. *et al.* (1980) Glucose metabolism and acetylcholine synthesis in relation to neuronal activity in Alzheimer's disease. *Lancet* **i**, 333

Sjöstrom, R. and Roos, B. E. (1972) 5-hydroxyindole acetic acid and homovanillic acid in cerebrospinal fluid of manic depressive psychosis. *Eur J Clin Pharmac* **4**, 170

Smith, C. M. and Swash, M. (1979) Physostigmine in Alzheimer's disease. *Lancet* **i**, 42

Snodgrass, S. R. (1978) Use of [^{3}H]-muscimol for GABA receptor studies. *Nature* **273**, 392

Snyder, S. H. (1977) Opiate receptors and internal opiates. *Scientific Am* **236**, 44

Snyder, S. H. (1980) Phencyclidine. *Nature* **286**, 355

Snyder, S. H., Creese, I. and Burt, D. R. (1977) Dopamine receptor binding in mammalian brain: relevance to psychiatry. In: *Neuroregulators and Psychiatric Disorders*, p 526 (Eds Usdin, E., Hamburg, D. A. and Barchas, J. D.). Oxford Univ. Press, New York

Snyder, S. H. and Taylor, K. M. (1972) Histamine in the brain: a neurotransmitter? In: *Perspectives in Neuropharmacology: a Tribute to Julius Axelrod*, p 44 (Ed S. H. Snyder) Oxford University Press, New York

Soubrie, P., Montastrac, J. L., Bourgoin, S., Reisine, T., Artaud, F. and Glowinski, J. (1981) In vivo evidence for GABAergic control of serotonin release in the cat substantic nigra. *Eur J Pharmac* **69**, 483

Spitzer, R. C., Endicott, J., Robins, E., Kunansky, J. and Gurland, B. (1975) Preliminary report of the reliability of research diagnostic criteria applied to psychiatric case records. In: *Predictability in Psychopharmacology, Preclinical and Clinical Conclusions* (Ed Sudilovsky, A., Gershon, S. and Beer, B.) Raven Press, New York

Squires, R. F. and Braestrup, C. (1977) Benzodiazepine receptors in rat brain. *Nature* **266**, 732

Stein, L., Wise, D. C. and Belluzi, J. D. (1975) Effects of benzodiazepines on central serotonergic mechanisms. *Adv Biochem Psychopharmac* **14**, 29

Task Force (1980) Task force on nomenclature and statistics of the American Psychiatric Association. *Diagnostic and Statistical Manual of Mental Disorders* American Psychiatric Association, Washington D.C.

Terenius, L. (1973) Characteristics of the 'receptor' for narcotic analgesics in synaptic plasma membrane fractions from rat brain. *Acta Pharmac Toxic* **33**, 377

Toffano, G., Guidotti, A. and Costa, E. (1978) Purification of an endogenous protein inhibitor of the high affinity binding of gamma-aminobutyric acid to synaptic membranes of rat brain. *Proc Nat Acad Sci* **75**, 4024

Tomlinson, B. E. (1980) The structural and quantative aspects of the dementias. In: *Biochemistry of Dementia* p 15 (Ed Roberts, P. J.) John Wiley & Sons, Chichester

Träskman, L., Åsberg, M., Bertilsson, L. and Sjistrand, L. (1981) Monoamine metabolites in cerebrospinal fluid and suicidal behaviour. *Archives of General Psychiatry* (in press)

Tregear, G. W. *et al.* (1971) Synthesis of substance P. *Nature (New Biol)* **232**, 87

Tuomisto, J. and Tukianen, E. (1976) Decreased uptake of 5-hydroxytryptamine in blood platelets for depressed patients. *Nature* **262**, 596

Twycross, R. (1977) Value of cocaine in opiate-containing elixirs. *Br med J* **2** 1348

Tyre, N. C., Iversen, S. D. and Green, A. R. (1979) The effects of benzodiazepines and serotonergic manipulations on punished responding. *Neuropharmacology* **18**, 689

Tyrer, P. (1974) The benzodiazepine bonanza. *Lancet* **ii**, 709

Tyrrell, D. A. J., Parry, R. P., Crow, T. J., Johnstone, E. C. and Ferrier, I. M. (1979) Possible virus in schizophrenics and some neurological disorders. *Lancet* **i**, 839

U'Pritchard, D. C., Greenberg, D. A., Sheenam, P. P. and Snyder, S. H. (1978) Tricyclic antidepressants: therapeutic properties and affinity for α-noradrenergic receptor binding sites in the brain. *Science* **199**, 197

Usdin, E. (1974) Neuropsychopharmacology of monoamines and their regulatory enzymes. *Adv Biochem Psychopharmac* **12** .

Uzan, A., Kabouche, M., Rataud, J. and Lefur, G. (1980) Pharmacological evidence of a possible tryptaminergic regulation of opiate receptors by using indalpine, a selective 5-HT uptake inhibitor. *Neuropharmacology* **19**, 1075

Van Praag, H. M. (1976) *Depression and Schizophrenia*. Spectrum Publ, New York

Ventulani, J., Stawarz, R. J., Dingell, J. V. and Sulser, F. (1976) A possible common mechanism of action of antidepressant drugs. *Naunyn Schmiedebergs Arch Pharmak* **293**, 109

von Euler, U. S. (1946) Aspecific sympathomimetic ergone in adrenergic nerve fibres (sympathin) and its relations to adrenaline and noradrenaline. *Acta physiol Scand* **12**, 73

von Euler, U. S. and Gaddum, J. H. (1931) An unidentified depressor substance in certain tissue extracts. *J Physiol (Lond)* **72**, 74

Weber, E. (1954) Ein Rauwolfia alkaloid in der Psychiatrie: seine Wirkungahnlichkeit mit Chlorpromazine. *Schweiz med Wschr* **84**, 968

Weinstock, M. (1980) Behavioural effects of β-adrenoceptor antagonists associated with blockade of central serotoninergic systems. In: *Enzymes and Neurotransmitters in Mental Disease* p 431 (Eds Usdin, E., Sourkes, T. L. and Youdim, M. B. H.) John Wiley & Sons, Chichester

Weissman, M. M. and Slaby, A. E. (1973) Oral contraceptives and psychiatric disturbance: evidence from research. *Br J Psychiatry* **123**, 513

West, E. D. (1981) Electro-convulsion therapy in depression: a double-blind controlled trial. *Br Med J* **282**, 355

Wilson, I. B. (1958) A specific antidote for nerve gas and insecticide (alkylphosphate) intoxication. *Neurology* (supp 1) **8**, 41

Wilson, I. L., Prange, A. J. and McClure, J. K. (1970) Thyroid hormone enhancement of imipramine in non-retarded depressions. *New Engl J Med* **282**, 1063

Wing, J. K., Cooper, J. E. and Sartorius, N. (1974) *The Description and Classification of Psychiatric Symptoms*. Cambridge Univ Press, London

World Health Organization (1973) *The International Pilot Study of Schizophrenia*, Vol I. Geneva

Wolstenholme, G. E. W. and Knight, J. (1976) *Monoamine Oxidase and its Inhibition*. Ciba Foundation Symposium **39** (New Series) Elsevier, Amsterdam

Woolley, D. W. and Shaw, E. (1954) A biological and pharmacological suggestion about certain mental disorders. *Proc Nat Acad Sci USA* **40**, 228

Wyatt, R. J., Portnoy, B., Kupfer, D. J., Snyder, F. and Engelman, K. (1971) Resting plasma catecholamine concentrations in patients with depression and anxiety. *Arch Gen Psychiat* **24**, 65

Yamamura, H. I., Enna, S. J. and Kuhar, M. J. (1978) *Neurotransmitter Receptor Binding*. Raven Press, New York

Yates, C. M., Blackburn, J. A. et al. (1980) Clinical and biochemical studies in Alzheimer's Disease. In: *Biochemistry of Dementia*, p 185 (Ed: Roberts, P. J.). John Wiley & Sons, Chichester.

Yerkes, R. M. and Dodson (1908) The relation of strength of stimulus to rapidity of habit formation. *J Comp Neurol Psychol* **18**, 459

Yorkston, N. J., Zaki, S. A., Pitcher, D. R., Gruzelier, J. H., Hollander, D. and Sergeant, H. G. S. (1977) Propranolol as an adjunct to the treatment of schizophrenia. *Lancet* **ii**, 575

Youdim, M. B. H., Collins, G. G. S., Sandler, M., Bevan-Jones, A. B., Pare, C. M. B. and Nicholson, W. J. (1972) Human brain monoamine oxidase: multiple forms and selective inhibitors. *Nature* **236**, 225

Youdim, M. B. H., Green, A. R., Bloomfield, M. R., Mitchell, M. D., Heal, D. J. and Grahame-Smith, D. G. (1980) The effects of iron deficiency on biogenic monoamine biochemistry and function in rats. *Neuropharmacology* **19**, 259

Young, S. N. and Sourkes, T. L. (1977) Tryptophan in the central nervous system: regulation and significance. *Adv Neurochem* **2**, 133

Youngblood, W. W., Lipton, M. A. and Kizer, J. S. (1978) TRH-like immunoreactivity in urine, serum and extrahypothalamic brain: non-identity with synthetic pyroglu-hist-proNH$_2$ (TRH). *Brain Res* **151**, 99

Zeller, E. A. (1959) The role of amine oxidases in the destruction of catecholamines. *Pharmac Rev* **11**, 387

Index

Acetaldehyde, involvement in alcoholism, 158–159
Acetyl CoA, role in acetylcholine synthesis, 32
Acetylcholine,
  agonists and antagonists, 34
  assay methods, 31
  cannabis effect on, 161
  characteristics of receptors, 33–34
  Huntington's chorea and brain systems of, 142
  metabolism of, 32–33
  neuroleptics and, 127
  radioligands for, 30
  receptor antagonists, 34
  receptors in Alzheimer's disease, 136
  synthesis of, 31–32
  tricyclic antidepressants and, 81
Acetylcholinesterase,
  Alzheimer's disease and activity of, 134–135
  structure and characteristics of 32–33
Addiction, alcohol, mechanisms involved in, 158–160
Adenylate cyclase,
  activity following antidepressants, 80
  dopamine sensitive and neuroleptics, 119–120
  histamine-sensitive, 38
  lithium effects on, 91
  monoamine sensitive, 25, 27
Adrenaline,
  distribution of, 19
  isolation and identification, 19
α-Adrenoceptors,
  effects of drugs, 24
  tricyclic antidepressants effect on, 80
β-Adrenoceptor antagonists, anxiety and, 102
β-Adrenoceptors,
  antidepressants and, 80
  electroconvulsive shock and, 87

ADTN, schizophrenia and binding in post-mortem brains, 112
Aetiology,
  role in classification of psychiatric illness, 1
  theories in schizophrenia, 110
Affective disorders, definition of, 49
Affinity, definition of, 29
Age, effect on monoamines and their enzymes, 131–133
Agitation, as symptom of depression, 50
Akinesia,
  dopamine involvement in, 147
  Parkinson's disease and, 145–146
Albumin, see Plasma albumin
Alcohol,
  biochemical aspects of addiction, 158–160
  consumption related to price, 156
  effect of thyrotropin releasing hormone on sedative effect, 46
  metabolism of, 158
  physical effects of, 157
  social aspects of addiction, 156
  tolerance to, 157
  withdrawal phenomena, 157
Alcohol abuse,
  clinical aspects, 157–158
  treatment of, 158
Alcohol dehydrogenase, role in alcohol metabolism, 159
Aldehyde reductase,
  characteristics of, 16
  role in alcohol metabolism, 159
  role in catecholamine metabolism, 22–23
Allylglycine, as glutamic acid decarboxylase inhibitor, 36
Alzheimer's disease,
  biochemistry of, 130–136
  clinical aspects of, 129–130
  cognitive function and use of drugs, 137

206

Alzheimer's disease (*continued*)
  future approaches to treatment, 138
  5-hydroxytryptamine and, 133
  pathological changes, 130–131
  pharmacology of, 136–138
  prevalence, 129
Amantadine, Parkinson's disease and the
    use of, 153
γ-Amino butyric acid, *see* GABA
Aminooxyacetic acid, as GABA
    transaminase inhibitor, 36
Aminopeptiases, and peptide inactivation,
    44
Amitriptyline, pharmacology of 75–83
Amphetamine,
  behaviour and biochemical effects,
    119
  firing rate in brain regions following,
    125
  pharmacology of, 168–169
  physical and psychological effects,
    167–168
  schizophrenic like behaviour produced
    by, 119
Amphetamine psychosis,
  clinical aspects of, 168
  discussion of, 119
Animal models of psychiatric illness, 5
Antibodies, possible role in tolerance, 156
Anticholinergic activity, neuroleptics and,
    125
Anticholinergic drugs, use in Parkinson's
    disease, 149, 153
Anticonvulsants, barbiturate use as, 171
Antidepressants,
  efficacy of, 72
  potentiation of actions by tri-
    iodothyronine, 46
Anxiety,
  β-adrenoceptor antagonists use in, 102
  barbiturate use in, 102
  benzodiazepines use in, 96–102
  classification of, 95
  guidelines to abnormal states, 93–94
  meprobamate use in, 102
  monoamine oxidase inhibitors use in,
    103
  performance and, 93
  pharmacology of, 96–103
  plasma catecholamines and, 96
  presence in depression, 51
  somatic or physical components, 93–94
  symptomatology of, 93–94
  tricyclic antidepressants and, 102

Apomorphine, dose and dopamine receptor
    activation, 24
Appetite loss, as symptom of depression,
    51
APUD system, concept of, 40
Atropine,
  as acetylcholine antagonist, 34
  use in Parkinson's disease, 149
Auditory hallucinations, schizophrenia
    and, 105
Autoreceptors, *see* Presynaptic receptors

B_{max}, definition of, 29
Baclofen, use in Huntington's chorea,
    143
Barbiturates,
  anxiety and, 102
  metabolism of, 172
  pharmacology of, 171–172
  physical and psychological effects of,
    170–171
  structure of, 171
  tolerance and, 171
  withdrawal and, 171
Basal ganglia, dopamine and Parkinson's
    disease, 146–147
Behaviour,
  amphetamine and, 169
  5-hydroxytryptamine mediated, 17
  neuroleptics effect on, 118–119
  substance P effect on, 47
Behavioural approaches, use in obsessional
    disorders, 95–96
Behavioural models, electroconvulsive
    shock and, 83–87
Benzodiazepines,
  anxiety and, 96–102
  clinical use of, 96–102
  dependence and, 170
  GABA modulin and, 98–99
  metabolism and kinetics, 101
  radioligands for, 30
  receptor for, 98–100
  self-poisoning and, 96
  structures of, 97
  tolerance and, 170
  withdrawal phenomena, 170
Bethanechol, action of, 34
Bicuculline, as GABA antagonist, 36
Biochemical measures, problems of
    investigating changes in psychiatric
    patients, 3–5
Bipolar depression, versus unipolar, 53
Blood pressure, anxiety and, 94

Brain,
  monoamine concentrations in post-
    mortem brains from depressives,
    63–64
  monoamine oxidase in, 15–16
  post-mortem and biochemistry of
    schizophrenia, 111–114
  problems in use of post-mortem tissue, 3
  tricyclic binding to, 82
Brief Psychiatric Rating Scale,
    schizophrenia and use, 109
Bromocriptine, Parkinson's disease treated
    with, 152–153
α-Bungarotoxin, action of, 34
β-Bungarotoxin, as inhibitor of choline
    uptake, 32
Butyrophenones, structures and use in
    schizophrenia, 117
Butyrylcholinesterase, characteristics of,
    32

Cannabis,
  pharmacology of, 160–161
  physical and psychological effects, 160
  tolerance in users, 160
β-Carboline-3-carboxylate, ligand for
    benzodiazepine receptor, 99–100
β-Carbolines, alcohol addiction and, 159
Catatonia, schizophrenia and motor
    behaviour, 150
Catecholamine metabolites, depression
    and, 68–69
Catecholamines,
  Alzheimer's disease and brain
    concentrations of, 131–132
  control of synthesis and release, 23
  distribution of, 18–19
  isolation and identification, 18
  metabolism of, 22–23
  plasma concentrations and depression,
    69
  synthesis of, 19–22
  see also Adrenaline, Dopamine, and
    Noradrenaline
Catechol-o-methyltransferase,
  inhibitor of activity, 23
  role in catecholamine metabolism,
    22–23
  schizophrenia and, 111
CATEGO,
  diagnosis, schedule of, 106
  use in psychiatric classification, 23
Cerebral blood flow, Alzheimer's disease
    and, 137

Cerebrospinal fluid,
  depression and catecholamine
    metabolites in, 68–69
  dopamine metabolites in Parkinson's
    disease, 147
  monoamine metabolite concentrations
    during depression, 60–63
  monoamine metabolite concentrations
    during mania, 89
  significance of monoamine metabolite
    measurements, 62–63
  tryptophan concentrations during
    depression, 65
  use in examining biochemical changes
    in psychiatry, 4
Challenge tests, use in investigatory
    biochemical changes in psychiatry, 5
Cheese reaction,
  deprenil and, 74
  role of monoamine oxidase inhibitors,
    72
p-chloroamphetamine, effects on brain
    5-hydroxytryptamine, 13, 169
p-chlorophenylalanine,
  as inhibition of tryptophan-5-
    hydroxylase, 13
  tricyclic antidepressant efficacy
    following, 80
Chlordiazepoxide, anxiety and, 96–102
Chlorimipramine, pharmacology of, 75–83
Choline,
  Alzheimer's disease and administration
    of, 137
  synthesis of acetylcholine and, 31–32
  transport of, 31–32
Choline acetyltransferase,
  Alzheimer's disease and activity of, 134
  characteristics of, 32
Choline uptake,
  Alzheimer's disease and reduction of,
    134
  inhibitors of, 32
Cholinergic systems, Alzheimer's disease
    and, 134–137
Cholinomimetric drugs, Alzheimer's
    disease and use of, 137
Chromosomes, lysergic acid diethylamide
    and damage, 165
Citalopram, pharmacology of, 75–83
Citric acid cycle, involvement in GABA
    synthesis, 35–36
Classification of anxiety, 95
Classification of depression, 51–53
Classification systems, use in psychiatry, 2

Clonidine, as α_2-adrenoceptor agonist, 24
Clorgyline, use as monoamine oxidase
    inhibitor, 15
Clozapine,
    site of action, 125
    structure and use in schizophrenia, 117
Cocaine,
    anaesthesia and, 169
    noradrenaline uptake and, 79
    physical, psychological, and
        pharmacological effects, 169
Codeine, see Opiates
Coexistence,
    concept and examples of, 40
    substance P and, 47
Cognitive function, cholinomimetic drugs
        and Alzheimer's Disease, 137
Compliance, problems in schizophrenia,
    108
Conflict punishment model,
    benzodiazepines and, 97–98
    plasma concentrations during
        depression, 65
Corticosteroids, as inducers of tryptophan
        pyrrolase, 10
Corticotrophan-related hormones, as
        peptide transmitters, 48
Cross-tolerance, definition of, 155
Curare, action of, 34

Dale hypothesis, contradiction of, 40
Deanol, use in Huntington's chorea, 143
Decarboxylase, characteristics of, 14
Decarboxylase inhibitors, L-dopa in
        combination with, 150–151
Delusions,
    depression and, 50
    mania and, 88
    schizophrenia and, 105
Dementia,
    as feature of Parkinson's disease, 144
    Huntington's chorea and, 139
    see also Alzheimer's disease
Dementia praecox, see Schizophrenia
Dependence,
    amphetamines and, 168
    benzodiazepine and, 170
    cocaine and, 169
    definition of, 155
    mechanisms involved in, 155–156
    phenocyclidine, 167
Depot neuroleptics, use of, 116
Deprenil,
    cheese reaction and, 74

use as monoamine oxidase inhibitor, 15
Depression,
    as feature of Parkinson's disease, 145
    association with life events, 49
    biochemical aspects of, 57–71
    catecholamine metabolites in
        cerebrospinal fluid and, 68–69
    cerebrospinal fluid concentrations of
        tryptophan, 65
    classification according to family
        history, 55
    classification of, 51–56
    clinical features, 50–51
    conceptual issues of classification,
        51–52
    conclusions on pharmacology of, 87–88
    electroconvulsive therapy and, 83–87
    endogenous and reactive, 54–55
    epidemiology of, 49–50
    flurothyl use in, 86
    genetic aspects, 53–54
    in vivo studies on neurotransmitter
        function, 70–71
    lithium and, 83
    monoamine metabolite concentrations
        in cerebrospinal fluid, 60–63
    monoamine turnover during, 62
    oral contraceptives and, 65–67
    peptides use in, 88
    pharmacological aspects, 71–92
    physiological symptoms, 51
    plasma catecholamines and, 69
    plasma corticosteroid concentrations
        during, 65
    plasma tryptophan concentrations in,
        64–65
    platelet 5-hydroxytryptamine uptake,
        68
    platelet monoamine oxidase activity, 68
    post-mortem brain monoamine
        concentrations, 63–64
    primary versus secondary, 54
    psychiatric aspects, 49–57
    psychotic and neurotic, 54–55
    quantification of, 56
    rating scales, 56
    social conditions and, 51–52
    tricyclic antidepressants and, 75–83
    tryptophan pyrrolase activity in, 65
    tryptophan use and, 74–75
    type A versus type B, 54–55
    unipolar versus bipolar, 53
    urinary 5-hydroxyindoleacetic acid
        concentrations, 67

Desmethylimipramine, α_2-adrenoceptor
   changes following, 80
Dexamphetamine, see Amphetamine
Dextropropoxyphene, see Opiates
Diagnosis, problems of psychiatric
   assessment, 1–2
Diagnostic and Statistical Manual, use in
   psychiatric classification, 2
Diagnostic schedules, schizophrenia use in,
   106–108
Diazepam, anxiety and, 96–102
Dibenzazepine, structure and use in
   schizophrenia, 117
Diethyldithiocarbamate, as dopamine
   β-hydroxylase inhibitor, 21
Dihydroxyphenylacetic acid, as dopamine
   metabolite, 23
3,4-Dihydroxyphenylglycol, as
   noradrenaline metabolite, 22–23
Dihydroxyphenylserine, as noradrenaline
   precursor, 21
5,7-Dihydroxytryptamine, use in lesioning
   of 5-hydroxytryptamine pathways, 8
Di-isopropyl flurophosphate, action of, 33
Diphenylbutylpiperidines, structure and
   use in schizophrenia, 116
Disulfiram,
   alchohol abuse treated with, 158
   dopamine β-hydroxylase inhibition and,
   21
Diurnal variation in depression, 50
L-Dopa,
   combination with peripheral
   decarboxylase inhibition,
   150–151
   formation from tyrosine, 21
   Parkinson's disease and use of,
   149–152
   pyridoxine in combination with, 151
   side effects, 150
   value in the treatment of Parkinson's
   disease, 152
L-Dopa decarboxylase,
   Alzheimer's disease and activity of, 131
   characteristics of, 14
   role in catecholamine formation, 21
L-Dopa potentiation test, for examining
   possible antidepressants, 46
Dopamine,
   adenylate cyclase stimulated by, 25–27
   adenylate cyclase stimulated by, and
   neuroleptics, 119–120
   behavioural responses in alcohol treated
   rats, 159

brain concentration in schizophrenia,
   111
brain concentrations after L-dopa
   therapy, 150
cannabis effect on, 161
concentrations in spinal cord, 68
distribution of, 18–19
interaction with GABA, 37
isolation and identification, 18–19
metabolism of, 22–23
Parkinson's disease and, 146–147
radioligands for, 30
receptor after long term neuroleptics,
   126
receptor supersensitivity after
   neuroleptics, 126
release by amphetamine, 168
synthesis of, 19–21
see also Catecholamines
Dopamine β-hydroxylase,
   Alzheimer's disease and, 133
   characteristics of, 21
   inhibitors of activity, 21
   schizophrenia and, 111
Dopamine metabolism, neuroleptics effect
   on, 118–119
Dopamine uptake, nomifensine and, 77
Drug abuse, definition of, 155
Drug dependency, general and social
   aspects of, 155–156
DSM-III, see Diagnostic and Statistical
   Manual
Dyskinesia, Parkinson's disease and,
   145–146

Edrophonium, action of, 33
Efficacy of antidepressants, 72
Electrochemical detectors, use in
   examining monoamine release, 17
Electroconvulsive shock,
   β-adrenoceptor binding and, 50, 87
   behavioural models and, 83–87
   effect on monoamines, 84–87
   noradrenaline sensitive adenylate
   cyclase and, 80
   see also Electroconvulsive therapy
Electroconvulsive therapy,
   criteria for successful treatments, 85–86
   depression and, 83–87
   Parkinson's disease treated with, 153
   pharmacology of, 83–87
   see also Electroconvulsive shock
End product inhibition, role in
   catecholamine metabolism, 23

β-Endorphin, pharmacology of, 42
Enkephalins,
   distribution of, 43–44
   involvement with 5-hydroxytryptamine, 44
   neuroleptics and, 128
   radioligands for, 30
   receptor types, 43
   role in opiate withdrawal, 163–164
   structures of, 42
   synthesis and degradation, 44
Enkephalitis lethargica,
   Parkinsonism and, 144
   prevalence, 145
Epidemiology, of depression, 49–50
Epilepsy, see Seizure disorders
Epinephrine, see Adrenaline
Erogt, role in St. Anthony's fire, 164
Eserine, as acetylcholinesterase inhibitor, 33
Extrapyramidal side effects, neuroleptics and, 115, 119

Family, history, classification scheme for depression, 55
Feighner criteria, schizophrenia use in, 107
First rank symptoms, schizophrenia and, 104–105
FLA-63, as dopamine β-hydroxylase inhibition, 21
Flashback, phenomenon of, in lysergic acid diethylamide abusers, 165
Fluoxetine, pharmacology of, 75–83
Flupenthixol, isomers and clinical efficacy, 123–124
   structure and use in schizophrenia, 116
Flurothyl, depression and, 86
Foods, monoamine oxidase inhibitors and, 72

GABA,
   agonists and antagonists, 36, 154
   alcohol addiction and involvement of, 159
   Alzheimer's disease and, 133
   anxiety and, 100–101
   Huntington's chorea and, 140–141
   interaction with dopamine, 37
   method of measuring turnover, 37
   neuroleptics and, 127
   Parkinson's disease and, 148–149
   radioligands for, 30
   synthesis of, 35–37

GABA mimetics, Parkinson's disease treated with, 154
GABA modulin, pharmacology of, 98–99
GABA transaminase, involvement in GABA metabolism, 36
Gallamine, action of, 34
Genetic aspects of depression, 53–54
Genetic counselling, Huntington's chorea and, 139
Genetic inheritance, Huntington's chorea and, 139
Glutamic acid,
   as GABA precursor, 36
   as neurotransmitter, 38
Glutamic acid decarboxylase,
   Alzheimer's disease and, 133
   Huntington's chorea and activity of, 140–141
   involvement in GABA synthesis, 36
   Parkinson's disease and activity of, 148–149
   schizophrenia and activity of, 114
Glycine, as neurotransmitter, 38
Growth hormone,
   measurement in depressives, 70–71
   use in challenge tests, 5

Hallucinogens,
   groups of compounds producing hallucination, 166
   historical and social aspects, 164
Halo effect, 3
Haloperidol, structure and use in schizophrenia, 116–117
Hamilton Rating Scale, general aspects, 56
Heart rate, anxiety and, 94
Hemicholinium-3, as inhibition of choline uptake, 32
Heroin, see Opiates
Hexamethonium, action of, 34
High affinity uptake, of choline, 31–32
Histamine,
   adenylate cyclase and antidepressants, 81
   neurotransmitter role of, 38
   radioligands for, 30
   synthesis and metabolism, 38–39
Homovanillic acid,
   as dopamine metabolite, 23
   brain concentrations after L-dopa therapy, 150
   cerebrospinal fluid concentrations during depression, 68–69
   Parkinson's disease and changes of, 146–147

Huntington's chorea,
  biochemistry of, 139–143
  catecholamine concentration and
    enzymes in, 141
  cholinergic systems in, 142
  clinical features, 139
  GABA systems and, 140–141
  5-hydroxytryptamine and, 142
  pathological changes in, 139
  pharmacology of, 143
Hydergine, use in Alzheimer's disease, 137
5-Hydroxyindole acetaldehyde,
  metabolism of, 16
5-Hydroxyindoleacetic acid,
  concentration in cerebrospinal fluid as
    predictor of suicide
    predisposition, 60–63
  concentration in cerebrospinal fluid of
    depressives, 60–63
  formation of, 14–16
  origin and significance in cerebrospinal
    fluid, 62
  urinary concentrations during
    depression, 67
5-Hydroxytryptamine,
  adenylate cyclase activity, 27
  Alzheimer's disease and brain
    concentrations of, 133
  anxiety and possible role of, 98,
    100–101
  cannabis effect on, 161
  control of synthesis, metabolism, and
    function, 16–18
  distribution of, 8
  effect of intraneuronal metabolism on
    function, 17
  effect of tryptophan availability on
    synthesis rate, 17
  effect on animal behaviour, 17
  Huntington's chorea and brain
    concentrations of, 142
  identification of, 8
  involvement with enkephalins, 44
  isolation of, 8
  metabolism of, 14–16
  Parkinson's disease and concentrations
    of, 147
  propanolol action of, 128
  radioligands for, 30
  receptors after long-term neuroleptics,
    127
  schizophrenia and, 111
  synthesis of, 9
  turnover studies using probenecid, 62

  uptake into platelets of depressed
    patients, 68
5-hydroxytryptophan decarboxylase,
  characteristics of, 14
5-hydroxytryptophol, formation of, 16
Hyoscyamine, see Atropine
Hypertensive crisis, role of monoamine
  oxidase inhibitors, 72
Hypomania, distinction between
  hypomania and mania, 88

Imipramine, pharmacology of, 75–83
Indoleamines,
  depression and concentrations in
    cerebrospinal fluid, 60–63
  substitution and hallucinogens, 166
  see also 5-Hydroxytryptamine
Insulin, effect on uptake of amino acids
  into brain, 17
Insulin coma and schizophrenia, 114
Intellect, in Alzheimer's disease, 129–130
Inter-rater reliability, 3
Iprindole, pharmacology of, 75–83
Iproniazid, effects on mood, 59
Isoniazid, effects on mood, 59

K_D, definition of, 29
Krebs cycle, involvement in GABA
  synthesis, 35–36
Kyrurenine, urinary concentration during
  depression, 65

Lecithin,
  acetylcholine synthesis and, 31
  Alzheimer's disease and administration
    of, 137
Letazol, see Pentylenetetrazol
Leucocytes, use in investigating
  biochemical change in psychiatry, 4
Levallorphan, as opiate antagonist, 163
Libido loss, as symptom of depression, 51
Life events, involvement in depression, 49
Ligand receptor binding,
  criteria for specific binding, 28–29
  effect of antidepressants on, 80
  neuroleptics and dopamine, 121–124
  schizophrenia and use of, 112–113
  technique of, 27–30
  use of techniques to investigate
    biochemical changes in
    psychiatry, 4
Lipotropin, structure of, 43
Lithium,
  depression and, 83

Lithium (*continued*)
　historical aspects of, 90
　measurement in plasma of, 90
　monoamine metabolism following, 83
　monoamine oxidase inhibitors and, 83
　pharmacology of, 83, 91–92
Lumbar puncture, use in psychiatric
　research, 62
Lysergic acid diethylamide,
　chromosome damage and, 165
　dopamine involvement in action of, 166
　5-hydroxytryptamine and, 165–166
　lack of withdrawal syndrome, 165
　pharmacology of, 165–166
　physical and psychological effects,
　　164–165
　structural similarity to 5-hydroxy-
　　tryptamine, 8, 58
　see also Hallucinogens

Major tranquillisers, *see* Neuroleptics
Mania,
　alternation with depression, 51
　biochemistry of, 89
　cerebrospinal fluid monoamine
　　concentrations and, 89
　pharmacology of, 89–92
　phenothiazines use in, 89
　psychiatric aspects, 88–89
Maprotiline, pharmacology of, 75–83
Marijuana, *see* Cannabis
Masked depression, definition of, 50
Melanotropin inhibitory factor, *see* Prolyl
　leucine glycine amide
Memory,
　Alzheimer's disease and, 130
　effects of vasopressin on, 48
Meprobamate, anxiety and, 102
Mescaline, *see* Hallucinogens
Mesolimbic forebrain, neuroleptics acting
　at this site, 124–125
Methamphetamine, *see* Amphetamine
α-Methyldopa,
　decarboxylase inhibition and, 14
　metabolism to α-methylnoradrenaline, 21
α-Methyldopahydrazine, pharmacology of,
　151
5-Methyltetrahydrofolate, role in forming
　β-carbolines, 159
α-Methyl-*p*-tryosine,
　as catecholamine synthesis inhibitor, 21
　effects on mania, 92
　tricyclic antidepressant efficacy
　　following, 80

Metoclopramide, non-antipsychotic action,
　125
Mianserin, pharmacology of, 75–83
Monoamine oxidase,
　Alzheimer's disease and, 133
　characteristics and distribution, 14–16
　inhibitors of, 15–16
　role in catecholamine metabolism,
　　22–23
　schizophrenia and activity of, 112
　substrates for, 15
Monoamine oxidase inhibitors
　anxiety and, 103
　characteristics of, 15–16
　combination with tryptophan, 74
　degree of inhibition produced by, 16, 74
　food induced interactions, 72
　lithium and, 83
　Parkinson's disease treated with,
　　151–152
　structure of, 73
　use in depression, 71–74
Monoamine synthesis, tricyclic
　antidepressants and, 77
Monoamine turnover, use of probeneid, 62
Monoamine uptake,
　effect of tricyclid antidepressants,
　　75–83
　selective inhibitors and, 77–83
Monoamines,
　electroconvulsive shock effect on,
　　83–87
　lithium effect on, 83
　measurement of turnover, 25
　metabolite concentration in
　　cerebrospinal fluid, 60–63
　methylated in schizophrenia, 110–111
　schizophrenia and, 111–114
Mood change, as feature of depression, 50
Morphine, *see* Opiates
Motor activity,
　depression and, 50
　schizophrenia and, 105
Muscarinic receptors, definition of, 33
Muscimol, as GABA agonist, 36
Muscle tension, anxiety and, 94

Naloxone, as opiate antagonist, 163
Negative features, condition of
　schizophrenia, 106
Negative self-concept, as symptom of
　depression, 50
Neostigmine, action of, 33

Neuroendocrine test, use in investigating biochemical changes in psychiatry, 5
Neurofibrillary tangles, Alzheimer's disease and, 130–131
Neuroleptics,
    depot preparations, 116
    dopamine sensitive adenylate cyclase activity following, 119–120
    effects of dopamine metabolism and behaviour, 118–119
    Huntington's chorea and use of, 143
    ligand binding studies and, 121–124
    long-term actions and schizophrenia, 126
    long-term administration and biochemical and behavioural consequences, 126
    mesolimbic forebrain and action of, 124–125
    receptor sensitivity and, 126
    theories of anti-schizophrenic action of, 127–128
    types of drugs and use in schizophrenia, 114–118
Neuromuscular blocker, types of, 34
Neuronal feedback loop, role in catecholamine synthesis regulation, 24
Neurotransmitter, criteria for, 35
Neurotransmitter function, *in vivo*, studies on, 70–71
Nicotine,
    action on acetylcholine receptors, 34
    mechanisms of dependence, 172–173
    physical and psychological effects, 172
    tolerance and withdrawal, 172
Nicotine receptors, definition of, 33
Nitrazepam, anxiety and, 96–102
Nomifensine, pharmacology of, 75–83
Non-esterified fatty acids, effect on plasma and brain tryptophan, 17
Noradrenaline,
    adenylate cyclase activity, 25–27
    adenylate cyclase activity and antidepressants, 80
    brain content in schizophrenia, 111
    distribution of, 19
    isolation and identification, 19
    metabolism of, 22–23
    radioligands for, 30
    synthesis of, 19–21
    see also Catecholamines
Norepinephrine, *see* Noradrenaline
Nortriptyline, dose and efficacy, 82

Nucleus accumbens, neuroleptics acting at this site, 124–125

Obsessional disorders, treatment of, 95–96
On–off phenomenon, clinical features of, 145–146
Operational criteria, in diagnosis of schizophrenia, 107
Opiate peptides, 41–44
Opiates,
    dependence on, 163–164
    historical aspects, 161–162
    physical and psychological effects, 162–163
    tolerance and withdrawal, 163–164
Opium, *see* Opiates
Oral contraceptives,
    effect on Vitamin B_6, 67
    tryptophan metabolism and depression, 65–67
Outcome, assessment in schizophrenia, 109–110
Oxygen, hyperbaric and Alzheimer's disease, 137

Pain, involvement of substance, 47
Panic attacks, 95
Pargyline, as monoamine oxidase inhibitor, 15
Parkinsonism,
    definition of, 144
    neuroleptics producing, 119
    see also Parkinson's disease
Parkinson's disease,
    amantadine in the treatment of, 153
    anticholinergics in the treatment of, 149, 153
    biochemistry of, 146–149
    bromocriptine in treatment of, 152–153
    clinical features of, 144–146
    L-dopa in the treatment of, 149–152
    dopamine changes in, 146–147
    electroconvulsive therapy in the treatment of, 153
    GABA concentrations in, 148–149
    GABA mimetics in the treatment of, 154
    5-hydroxytryptamine concentrations in, 147
    monoamine oxidase inhibitors in the treatment of, 151–152
    natural history of, 145
    new approaches to treatment, 153
    pharmacology of, 149–154

Parkinson's disease (*continued*)
  prevalence, 145
  prolyl leucine glycine amide in
    treatment of, 154
  stereotaxic surgery in the treatment of,
    149
Pentobarbital, effecto of thyrotropin
    releasing hormone on sedative effect,
    . 46
Pentylenetetrazol, as GABA antagonist, 36
Peptides,
  depression and use of, 88
  historical aspects 39–41
  list of possible transmitter candidates,
    41
Perceptual disorders, schizophrenia and,
    105
Performance, anxiety and, 93
Peripheral decarboxylase inhibitors, uses
    of, 14
Periqueductal grey,
  involvement in analgesia, 42, 44
  involvement with opiate peptides, 42
Personality, in Alzheimer's disease, 130
Phencyclidine,
  pharmacology of, 167
  physical and psychological effects, 166
Phenelzine, as monoamine oxidase
    inhibitor, 15
Phenothiazines,
  schizophrenia use in, 114–118
  subclasses of, 115
  structures of, 115
Phenylalanine, as catecholamine precursor,
    19
Phenylalanine hydroxylase, role in tyrosine
    formation, 21
Phenylethanolamine-*N*-methyl transferase,
    characteristics of, 22
Phenylethylamines, hallucinogens and, 166
Phobic anxiety, 95
Phosphatidyl choline, *see* Lecithin
Physostigmine,
  acetylcholinesterase inhibition and, 33
  Alzheimer's disease and use of, 137
Picrotoxin, as GABA antagonist, 36
Pilocarpine, action of, 34
Pimozide, structure and use in
    schizophrenia, 117
Pineal, monoamine oxidase in, 15
Pink spot, schizophrenia and, 111
Piracetam, use in Alzheimer's disease, 137
Plaque formation, Alzheimer's disease and,
    130–131

Plaques, choline acetyltransferase activity
    and number of, 135
Plasma albumin, binding with tryptophan, 9
Plasma tryptophan,
  as determinant of brain tryptophan
    concentration, 10–12
  effects of non-esterified fatty acid
    concentration, 17
  free and bound ratio, 9–10
Platelets,
  catecholamines and depression, 69
  5-hydroxytryptamine uptake during
    depression, 68
  model of the serotoninergic neurone, 68
  monoamine oxidase activity during
    depression, 68
  schizophrenia and activity of
    monoamine oxidase, 112
  tricyclic binding to, 82
  use in investigating biochemical changes
    in psychiatry, 4
Positive features, condition of
    schizophrenia, 106
Post-mortem brains, use in investigating
    biochemical changes, 3
Practolol, anxiety and, 102
Prazocin, as α_1-adrenoceptor antagonist, 24
Present state examination,
  diagnostic schedule, 106
  use in psychiatric classification, 2
Presynaptic receptors, role in controlling
    monoamine function, 23–24
Price, factor in alcohol consumption, 156
Probenecid,
  measurement of monoamine turnover in
    depression using, 62, 69
  measurement of monoamine turnover
    studies in mania using, 89
  use in measuring turnover, 25, 62
Prolactin,
  measurement in depressives, 70
  neuroleptics and plasma levels of, 119
  use in challenge tests, 5
Prolyl leucine glycine amide,
  behavioural effects of, 47
  Parkinson's disease treated with, 154
Pro-opiocortin, as opiate peptide precursor,
    44
Propanolol,
  5-hydroxytryptamine behaviour and
    biochemistry following, 128
  schizophrenia use in, 128
Protein phosphorylation, mechanisms
    involved in, 25–27

PSE, *see* Present state examination
Psilocybin, *see* Hallucinogens
Psychiatric assessment, problems of
  diagnosis, 1–2
Psychiatric illness,
  animal models, 5
  classification systems, 1–2
  validity of concept, 1
Psychoactive drugs, problems in investigat-
  ing their pharmacology, 5
Psychopharmacology, value of research, 7
Psychotogens, endogenous in
  schizophrenia, 110–111
Pyridine-2-aldoxime, as antidote to ace-
  tylcholinesterase inhibition, 33
Pyridoxine, L-dopa in combination with,
  151
Pyrogallol, as inhibitor of catecholamine-
  O-methyl transferase, 23

Quantal release, theory of, 34

Radioligands, use of, 27–30
Raphe region, as part of 5-hydroxy-
  tryptamine system, 8
Rating scales,
  for depression, 56
  problems of use, 3
Receptors,
  general aspects of, 27
  radioligands for, 30
Research diagnostic criteria, use in psychiatric classification, 2
Reserpine,
  effect on mood, 59
  schizophrenia use in, 114
  sedative effect of, 60
Respiration, anxiety and, 94
Retardation, as symptom of depression, 50
Re-uptake, monoamine inactivation and,
  16, 75
Rigidity,
  dopamine involvement in, 147
  Parkinson's disease and, 144

Saint Anthony's fire, reasons for, 164
Scatchard analysis, method of, 29
Schizo-affective disorders, psychiatric
  aspects of, 106
Schizophrenia,
  binding of spiropendol in post-mortem
    brains, 112–113

biochemistry of, 110–114
clinical features, 104–106
concept of, 104
diagnosis of, 106
monoamine receptor changes in,
  112–113
monoamines and, 111–114
neuroleptic actions after long-term
  treatment, 126
outcome of, 108–109
pharmacology of, 114–128
pink spot and, 111
propranolol use in, 128
psychiatric aspects of, 104–110
virus involvement, in 114
Second messenger, 25–27
Seizure disorders,
  benzodiazepines and, 102
  involvement with GABA, 37
Seizure threshold, benzodiazepines and,
  99–100
Self-rating scale, use in depression, 57
Senile dementia, *see* Alzheimer's disease
Serotonin, *see* 5-Hydroxytryptamine
Seryltrihydroxybenzylhydrazine, pharmacology of, 151
Side effects, L-dopa therapy and, 150–151
Simple schizophrenia, 106
Skin electrical conductance, anxiety and,
  94
Slang names,
  amphetamines, 168
  barbiturates, 171
  cannabis, 160
  cocaine, 169
  hallucinogens, 164
  opiates, 162
  phencyclidine, 167
Sleep disturbance, as symptom of depres-
  sion, 51
Smiling depression, definition of, 50
Sodium valproate, use in Huntington's
  chorea, 143
Spinal cord, dopamine in, 68
Spiperone, *see* Spiroperidol
Spiroperidol,
  Alzheimer's disease and binding in post-
    mortem brain, 132
  schizophrenia and binding in post-
    mortem brain, 112–113
Spitzer criteria, schizophrenia use in, 107
Stereotaxic surgery, Parkinson's disease
  use in, 149
Strychnine, as glucine antagonist, 38

Substance P,
  behavioural effects of, 47
  identification and localization, 46–47
  involvement in pain, 47
Succinylcholine, as neuromuscular blocker, 34
Suicide,
  Huntington's chorea and, 139
  monoamine metabolite concentrations in cerebrospinal fluid, 60–63
  possible consequence of depressive illness, 64
  predictive value of 5-hydroxy-indoleacetic acid in cerebrospinal fluid, 60–63
Sulpiride, site of action, 125
Swinging, clinical features of, 145–146

Tardive dyskinesia, clinical and biochemical aspects, 125–126
Tetrabenazine, use in Huntington's chorea, 143
Δ^9-tetrahydrocannabinol, see Cannabis
Tetraisoquinolines, alcoholism and their role in, 159
Therapeutic window, tricyclics and, 82
Thioridazine, anticholinergic activity, 125
Thioxanthines, structures and use in schizophrenia, 115
Thought disorders, schizophrenia and, 105
Thyrotropin releasing hormone,
  behavioural effects of, 46
  distribution and synthesis of, 44–46
  release of dopamine and, 46
Tiredness, as symptom of depression, 51
Tobacco, see Nicotine
Tolerance,
  barbiturates and, 171
  benzodiazepines and, 170
  cocaine and, 169
  definition of, 155
  mechanisms involved, 155–156
  nicotine and, 172
  opiates and, 163
Tranylcypromine,
  amphetamine-like actions of, 73
  monoamine oxidase inhibition and, 15
Tremor, Parkinson's disease and, 144
Tricyclic antidepressants,
  anxiety use in, 102
  binding to platelets and brain, 82
  depression and, 75–83
  development of, 60

efficacy and dose of, 82
  pharmacology of, 75–83
  structures of, 76
  time course of effects, 79
  tri-iodothyronine potentiates action of, 46
Tri-iodothyronine, potentiation of antidepressant drug actions, 46
Tropolones, as inhibitor of catechol-O-methyltransferase, 23
Tryptophan,
  brain metabolism, 11–12
  cerebrospinal fluid concentrations in depression, 65
  combination with a monoamine oxidase inhibitor, 74
  concentration in plasma and brain, 9–12
  concentration in plasma following various treatments, 17
  depression, use in, 74–75
  dose and efficacy, 75
  induction of tryptophan pyrrolase, 10
  metabolism after oral contraceptives, 65–67
  peripheral metabolism, 10
  plasma concentrations in depression, 64–65
  precursor of 5-hydroxytryptamine, 9–11
  transport systems into brain, 12
Tryptophan pyrrolase,
  activity in depression, 65
  inhibitors of, 75
  role in controlling tryptophan metabolism, 10
Tryptophan-5-hydroxylase, distribution and characteristics, 12–13
d-Tubocurarine, action of, 34
Turnover,
  concept of, 25
  techniques available and assumptions made, 25
Tyramine, hypertensive crisis and role of, 72
Tyrosine,
  as catecholamine precursor, 19
  transport into brain, 19
Tyrosine hydroxylase,
  Alzheimer's disease and activity of, 131
  distribution of, 21
  role in regulation of catecholamine synthesis, 23
  schizophrenia and, 111

Unipolar depression, versus bipolar, 53
Unreality feelings in depression, 50

Vanillyl mandelic acid, as noradrenaline
    metabolite, 22–23
Vasopressin,
    effects on memory, 48
    peptide transmitter candidate, 48
Viloxazine, pharmacology of, 75–83
Virus, schizophrenia and possible involve-
    ment of, 114
Visual analogue scale, use in depression, 57
Vitamin B$_6$, deficiency produced by oral
    contraceptives, 67

Withdrawal,
    alcohol and, 157

amphetamine and, 168
barbiturates and, 171
benzodiazepines and, 170
hallucinogens and, 165
nicotine and, 172
opiates and, 163–164
Withdrawal state, definition of, 155–156
World Health Organisation,
    drug abuse and dependence definitions
        of, 155
    schizophrenia diagnosis study, 106

Yohimbine, as α_2-adrenoceptor antagonist,
    24

Zimelidine, pharmacology of, 75–83